How to Test Almost Everything Electronic

2nd Edition

D1379633

How to Test Almost Everything Electronic

2nd Edition

Jack Darr and Delton T. Horn

TAB BOOKS
Blue Ridge Summit, PA

SECOND EDITION
SIXTH PRINTING

© 1988 by **TAB Books**.
TAB Books is a division of McGraw-Hill, Inc.

Library of Congress Cataloging-in-Publication Data

Darr, Jack.
    How to test almost everything electronic.

    Includes index.
    1. Electronic apparatus and appliances—Testing.
I. Horn, Delton T.   II. Title.
TK7878.D3   1988   621.381'028'7   87-35674
ISBN 0-8306-7925-1
ISBN 0-8306-2925-4 (pbk.)

TAB Books offers software for sale. For information and a catalog, please contact
TAB Software Department, Blue Ridge Summit, PA 17294-0850.

Questions regarding the content of this book should be addressed to:

Reader Inquiry Branch
TAB Books
Blue Ridge Summit, PA 17294-0850

OTHER BOOKS IN THE
TAB HOBBY ELECTRONICS SERIES

This series of five newly updated and revised books provides an excellent blend of theory, skills, and projects that lead the novice gently into the exciting arena of electronics. The Series is also an extremely useful reference set for libraries and intermediate and advanced electronics hobbyists, as well as an ideal text.

The first book in the set, *Basic Electronics Course—2nd Edition*, consists of straight theory and lays a firm foundation upon which the Series then builds. Practical skills are developed in the second and third books, *How to Read Electronic Circuit Diagrams—2nd Edition* and *How to Test Almost Anything Electronic—2nd Edition*, which cover the two most important and fundamental skills necessary for successful electronics experimentation. The final two volumes in the Series, *44 Power Supplies for Your Electronic Projects* and *Beyond the Transistor: 133 Electronics Projects*, present useful hands-on projects that range from simple half-wave rectifiers to sophisticated semiconductor devices utilizing ICs.

How to Read Electronic Circuit Diagrams—*2nd Edition*
ROBERT M. BROWN, PAUL LAWRENCE, and JAMES A. WHITSON
Here is the ideal introduction for every hobbyist, student, or experimenter who wants to learn how to read schematic diagrams of electronic circuits. The book begins with a look at some common electronic components; then some simple electronic circuits and more complicated solid-state devices are covered.

44 Power Supplies for Your Electronic Projects
ROBERT J. TRAISTER and JONATHAN L. MAYO
Electronics and computer hobbyists will not find a more practical book than this one. A quick, short, and thorough review of basic electronics is provided along with indispensable advice on laboratory techniques and how to locate and store components. The projects begin with simple circuits and progress to more complicated designs that include ICs and discrete components.

Beyond the Transistor: 133 Electronics Projects
RUFUS P. TURNER and BRINTON L. RUTHERFORD
Powerful integrated circuits—both digital and analog—are now readily available to electronic hobbyists and experimenters. And many of these

ICs are used in the exciting projects described in this new book: a dual LED flasher, an audible continuity checker, a proximity detector, a siren, a pendulum clock, a metronome, and a music box, just to name a few. For the novice, there is information on soldering, wiring, breadboarding, and troubleshooting, as well as where to find electronics parts.

Basic Electronics Course—2nd Edition
NORMAN H. CROWHURST

This book thoroughly explains the necessary fundamentals of electronics, such as electron flow, magnetic fields, resistance, voltage, and current.

Contents

Introduction

Electronics consists of things that we can neither hear, see, touch, nor taste. Electrical voltage, current, and resistance are invisible to the senses. A possible exception is high voltage. If you get an unwary fingertip into that, you can definitely *feel* it! However, that would be doing it the hard way. To learn what's going on in our circuits, you use test equipment—special apparatus designed to give us visible or aural indications of what's there and what it's doing.

Test instruments are amazingly versatile. They will literally do anything, if we know how to use them, how they work, and what their limitations are. It is just as important to know what a test instrument *can't* do as to know what it can. All instruments will make the tests for which they were designed; most will make many other tests as well, if we know how the instruments work and how to take advantage of their readings to indicate the presence or absence of other quantities.

That's what this book is about—electronic tests and measurements, how to make them with all kinds of electronic test equipment, and how to interpret the results. Interpretation is the most important point of the whole process. It requires a full knowledge of the test equipment *and* the circuits in which we're taking the readings.

Begin with the fundamentals. You might be familiar with these, but read the material anyway. You never can tell when you'll run into something new, or handy, that you hadn't thought of before. This happens to me often.

There are three basic groups of measurements: quantity—meaning voltage, current, and resistance, plus capacitance and inductance; output—meaning the normal output of a certain unit, stage, or device; and finally, quality—which answers the question, "Are the output and operation of this device up to normal?"

All test methods are combinations of these measurements, used as steps of a logical test sequence. Using the right methods and equipment and a logical sequence of tests, you will get results much sooner. For example, in testing a given stage, the first test should be qualitative: "Is the stage putting out the right amount and kind of signal?" If not, then test for quantities—the correct and proper operating voltages and currents. When you find something off-value in one of these, then test the parts supplying the voltage and current to that stage. Then find the defective one, replace it, and the trouble is fixed.

Dc voltage measurements are probably the most common tests on all kinds of electronic apparatus. Why? Simply because they are the fastest, easiest, and can tell you the most. By using them, you can get the maximum amount of information in a minimum amount of time. You don't even have to break the circuit—simply connect one lead of the voltmeter to a common or reference point and then check for the presence or absence of proper voltages with the other lead.

By the correct interpretation of dc voltage readings in a given circuit, you can tell if the circuit is drawing the proper current. In some cases, you can tell whether there is a short circuit to ground, and so on.

How can you check current by taking a *voltage* reading? First note the voltage present at a given point in the circuit. In almost every case, the current will have to go through at least one resistor to reach the test point. By noting whether the voltage at the test point is high, low, or normal, you can tell if the correct amount of current is flowing through the resistor. I go into more detail when going through the various tests for tubes, transistors, and other devices.

All voltage readings—in fact, all instrument readings of any kind—are comparisons. The value read must be compared against a standard before you can tell whether the test reading is high, low, or okay. For example, if we put a voltmeter on the plate of a tube and read 100 volts, this means absolutely nothing, but if the standard value given on the schematic diagram says the tube should have 200 volts, then you have learned something useful.

Experience is often very helpful. If you are familiar with the circuit of a typical TV set, for example, you already have a good idea of what voltages to expect on various other circuits. If you find the plate voltage of a resistance-coupled amplifier tube is too high, you know the tube isn't drawing enough current. The tube could be weak, the load resistor could have changed in value, or the grid bias could be too far negative.

If the collector voltage of a transistor output stage of a radio is too low, the collector current is too low. These deductions come from Ohm's law—current and voltage always have the same relation to each other no matter what kind of circuit you find them in. The most important difference between the tube and transistor circuits is in the magnitude of the voltages and currents: tubes use high voltages and low currents; transistors use low voltages and high currents. But in every case, you get your first clue to the cause of trouble by taking a set of voltage readings.

An important part of making tests is *accuracy*. Test equipment must be within a certain range of accuracy with respect to the standard. However, there are a lot of cases where absolute accuracy is not as important as many people think.

In tube circuits for instance, plate voltages have a wide tolerance: often as high as ±10 percent. This means that a tube requiring 100 volts on the plate can have any value between 90 and 110 volts and still be okay. In transistor circuits, however, with a voltage reading like "base bias with respect to emitter, 0.4 volt," accuracy must be considerably better. In this circuit, a voltage change of only 0.1 volt can cut off a transistor completely. You must know when to strive for extreme accuracy and when not to bother. This will be covered as we go through the various test procedures.

There are many shortcuts, combinations of instruments, etc., that you can use to test any quantity—even those that are apparently not in the range of such equipment. I give as many of these as are appropriate. There will be others that you can work out for yourself. A great many of these tests can be made with such complex and expensive equipment as a pilot lamp, a neon lamp or a dc voltmeter.

Jack Darr

In the twenty years since the first edition of *How to Test Almost Everything Electronic* was released, there have been a lot of changes in the field of electronics. Tube-based equipment was once the norm, but tubes are becoming increasingly rare, except in a few specialized applications. Many new types of semiconductor devices have been developed over the years. Probably the most significant development of the past two decades is the *integrated circuit,* or IC.

Obviously, new electronic devices call for new testing procedures, and this new edition focuses on the electronic equipment of today, while not completely ignoring older equipment that might still be in use.

The basic principles of troubleshooting and interpreting test results have not changed. Only the specifics of the testing procedures need to be altered to meet the requirements of today's electronic circuits. This edition also covers some new types of test equipment that have become available in recent years.

Finally, when the first edition of this book appeared back in 1966, few electronics technicians ever encountered digital circuitry. Today, of course, digital circuitry is everywhere. Testing a digital circuit is quite different from testing an analog circuit; hence, this new edition also features a section on digital testing.

Throughout, this book has attempted to be as generally useful as possible. It doesn't go into the specifics of servicing any given piece of equipment. Instead, you should examine the typical circuit types that are frequently encountered in various types of equipment. In updating the material in this book, I have tried to live up to the title and help the reader learn "how to test almost everything electronic."

Delton T. Horn

1
Test Equipment

In some cases you can diagnose a problem simply by logically examining the symptoms. More commonly, however, any given symptom or set of symptoms indicates several possible faults. The service technician depends on his test equipment. While certainly some technicians are better than others, it's true to an extent that no technician is better than his or her test equipment. The best technician in the world would be severely limited with inadequate test equipment.

Various common types of test equipment are mentioned throughout this book. In this section, I specifically focus on several different types of test equipment, emphasizing fairly recent devices as much as possible.

The first requirement for a well-equipped electronics workbench is a large work surface. You might find yourself with several pieces of equipment opened up and spread out at the same time, plus you need room for any test equipment. Therefore, your work surface should be as large as possible.

Adequate lighting is an absolute must. A dimly lit work area will slow you down and increase the chance of errors. Use multiple light sources so you can't block off the light with your own body or a piece of equipment.

A high-intensity lamp on a flexible goose-neck, as illustrated in Fig. 1-1, can be an extremely handy item to have—you can focus the light wherever you need it. A lot of equipment have dark little nooks and crannies you will need to see into.

Your workbench area should have several electrical sockets. You can never have too many. There always seems to be one more thing you need to plug in. Avoid using cube taps and extension cords as much as possible, because they can be fire hazards if overloaded. Many distributors sell power strips that are convenient and safe. Many even have built-in surge protectors. Most have a handy master power switch that is useful in some circumstances. For one thing, it is a good way to make sure everything is off when you close up for the night. Most commercial power strips look like the one in Fig. 1-2.

1

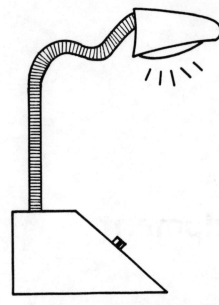

Fig. 1-1. A high-intensity lamp with a flexible goose-neck is a useful item on the workbench.

What test equipment will you need? To a large extent, this depends on just what type of work you do. A multimeter (VOM or VTVM) is absolutely essential on all electronics workbenches. If possible, you should have at least one VOM and one VTVM. The multimeter is the technician's right arm.

The technician's left arm is the oscilloscope. This is another essential device for virtually all electronics workbenches. Anytime you are dealing with an ac signal, a scope is useful, because it permits you to actually see the waveform. You can also measure the voltage at any point in the cycle, check for distortion, measure the cycle period, and many other tests.

The multimeter and the oscilloscope are the standard items and are used in all types of electronics work. Some other generally useful devices include variable power supplies, Variacs, and signal generators. There are many different types of signal generators for various applications; several common types are discussed in this chapter.

There are many specialized types of electronics test equipment. The usefulness of these devices depends on the specific type of work you do. These devices include capacitance meters, frequency counters, specialized signal generators, and signal tracers, among others.

Don't scrimp on your test equipment. Your accuracy will be no better than the capabilities of your equipment. Don't overspend on features you don't need, or even equipment you will rarely, if ever, use. But make sure you do get what you need. Nothing is worse than being stumped on a servicing job because you don't have the right equipment. Occasionally this is inevitable, because you will undoubtedly run into something outside your usual area every now and again. But if insufficient equipment problems hit you frequently, then you need to be better prepared.

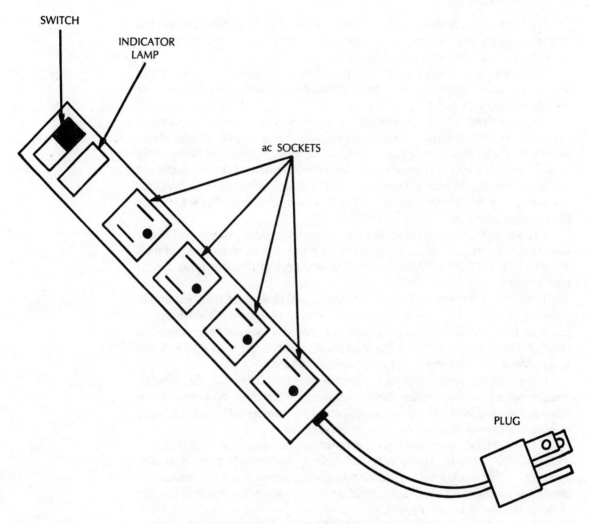

SWITCH

INDICATOR
LAMP

ac SOCKETS

PLUG

Fig. 1-2. A power strip safely provides extra ac outlets.

MULTIMETERS

Obviously, a multimeter is a meter that serves multiple functions. It measures various electrical parameters—usually voltage, current, and resistance. Multimeters have been used in many of the test procedures described in this book, especially in Chapters 2 and 4.

A VOM (volt-ohm-milliammeter) is a passive device. It does not include an active amplifier or buffer stage, so the input impedance (resistance) is fairly low. Some cheap VOMs have input impedances as low as 1000 ohms per volt. Such units are virtually useless for practical electronics work because they can excessively load down the circuit being tested, throwing off the measurement.

The unofficial standard for professional VOMs has long been considered 20,000 ohms-per-volt. Modern high-quality VOMs usually have input impedances of 50,000 or 100,000 ohms per volt.

VOMs are usually very light-weight and portable. The technician is not tied to a power source for the meter. Most VOMs have a small internal battery for the ohmmeter section.

In some precision measurements, a much higher meter impedance is necessary. In this case, use a multimeter with an active amplifier/buffer stage. In the past, the amplifier was a tube circuit, so this type of device has traditionally been known as a VTVM (vacuum tube voltmeter). Today, of course, tubes are becoming increasingly scarce. Even when such a device uses only semiconductor components, it is still often called a VTVM. A better name might be EMM, or Electronic Multimeter.

VTVMs, or EMMs, have high input impedances, typically one megohm per volt. They are very precise, but they tend to be more expensive and bulkier than VOMs. Also, a VTVM or EMM, being an active device, requires some sort of power supply.

Today the trend is toward digital. DMMs (digital multimeters) are extremely popular. Many of them are as small and portable as VOMs. Often they are powered from a small internal battery (9-volt transistor radio batteries are often used). A DMM usually has an input impedance comparable with a VTVM, but it also offers the convenience of a light-weight VOM.

Many digital multimeters offer special functions such as capacitance measurement, diode/transistor tests, frequency counters, or conductance measurements. (*Conductance* is the reciprocal of resistance (1/R) and is used to measure extremely small resistances accurately.)

A DMM displays the measured value directly in numerical form, for example, "12.73 volts." This is usually very convenient, and usually permits greater accuracy in the reading than a traditional analog meter. It is often difficult to determine precisely where the pointer is on a meter scale, especially if you are viewing the meter from an angle.

However, DMMs are not good for all applications. In some cases, the precise value will be less important than how the value changes over time. In such applications, a DMM will just give you a meaningless blur of rapidly changing numbers.

A well-stocked electronics workbench should have at least an analog VOM and a DMM. An analog VTVM or EMM would also be very desirable. If you can afford it, it is often useful to have several VOMs with spring-loaded clip leads. You can then simultaneously monitor different parts of a circuit.

ADAPTERS FOR MULTIMETERS

The standard multimeter measures volts, ohms, and milliamperes, so it is quite a versatile device. But it can be made even more versatile with special add-on adapters to permit additional types of measurements. Usually these adapters convert some other parameter into a proportional voltage.

These adapters need not be particularly complex. For example, an ordinary semiconductor diode can be used as a simple temperature-to-voltage converter. The scale won't be linear over a very wide range, but in some applications, it is sufficient.

A number of multimeter adapters are available commercially, especially from manufacturers of DMMs. Usually, however, you have to build one yourself. Plans can be found for these adapters in the popular electronics magazines (*Radio Electronics, Modern Electronics, Hands-On Electronics,* etc.).

Typical multimeter adapters create a proportional dc voltage from such parameters as

- temperature
- capacitance
- frequency
- light intensity

Other adapters extend the range of the multimeter, permitting you to read very large or very small voltages and/or currents. Adapters are also available to accurately measure ac voltages above the basic line frequency (60 Hz). For example, the circuit shown in Fig. 1-3 converts an ac voltage in the rf range into a proportionate dc voltage. Adapters for measuring true rms values by converting them to a proportionate dc voltage is discussed in Chapter 8.

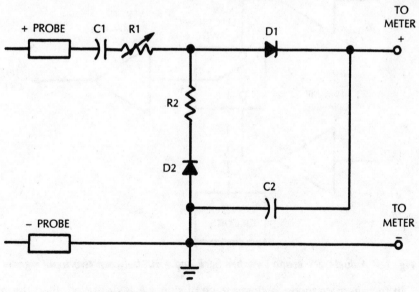

R1	10K TRIM POT (CALIBRATE—ADJUST FOR APPROXIMATELY 6.5K)
R2	47K RESISTOR
C1	0.82 μF CAPACITOR
C2	120 pF CAPACITOR
D1, D2	SMALL SIGNAL DIODES

Fig. 1-3. This circuit converts an ac voltage in the rf range to a proportionate dc voltage.

OSCILLOSCOPES

The oscilloscope is a powerful piece of test equipment that permits the technician to directly view the waveshapes of ac signals. This device has been used in many of the test procedures described in later chapters of this book, especially Chapters 5 and 7.

A couple of decades ago, most technicians had to get by with a single-trace scope. This is an oscilloscope that displays only one signal at a time. Dual-trace (two separate signals simultaneously displayed) capability was limited to expensive, laboratory-grade scopes.

Dual-trace scopes can be very helpful in servicing electronic equipment. An input signal and the resulting output signal can be simultaneously viewed and easily compared. Distortion and phase shift are clearly displayed.

In a dual-trace scope, a single vertical driver amplifier serves double duty for two separate preamp channels (inputs). The preamp feeding the driver amp at any given instant is determined by an electronic switch. This is illustrated in the block diagram of Fig. 1-4. The electronic switch can be operated in either of two modes; alternate and chopped.

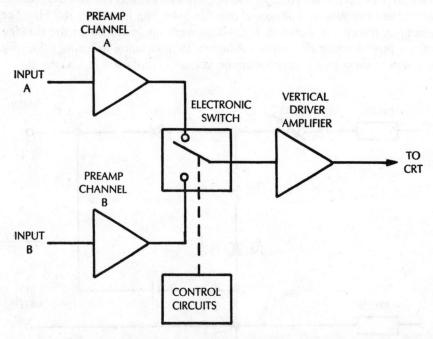

Fig. 1-4. A dual-trace scope switches back and forth between two input signals.

In the *alternate* mode, a single trace of signal A is displayed, then during the retrace, the electronic switch is reversed and a single trace of signal B is displayed. During the retrace of B, the electronic switch goes back to position A, and the process is repeated. The phosphor coating the CRT screen has enough persistence that it will glow longer than the sweep time. That is, before the image of one channel fades, the scope is ready to retrace it. Thus, both images appear on the screen at the same time.

To separate the images, the scope adds an offset voltage forcing the trace above or below the center line of the screen. With high-level signals at high gain, the traces might overlap. The solution is to reduce the scope's gain.

In the chopped mode, the electronic switch is operated at a very high rate—much higher than the sweep rate. A small segment of trace A is drawn, then a small segment of trace B, then A, then B, and so forth, until both traces are completed. Obviously, when the electronic switch is set up for channel A, part of signal B is lost from the trace. Each trace is made up of dashed lines; 50 percent of each trace is missing. However, the switching rate is so fast, that the gaps in the traces are extremely small and usually invisible. The traces appear to be made up of solid lines.

Dual-trace scopes are now the norm, except for small portable scopes and very inexpensive bench scopes. Single-trace scopes are still manufactured and used, but they are becoming increasingly rare. In the future, there will be even more extensive multiple-image scopes.

SIGNAL GENERATORS

A signal generator is used to provide a known input signal for signal-tracing type tests (like many of those described in Chapter 9). The standard procedure for signal tracing is to feed the signal generator's output to the input of the circuit to be tested. In some cases, it is necessary to disconnect the circuit's normal input connection. A measurement is then taken at the output of the first stage of the circuit. Depending on what is being tested, look to see if the signal is missing, overly attenuated, distorted, or clipped. If the signal looks okay at the output of the first stage, move on to the output of the second stage. Continue with this process, moving toward the circuit's final output.

When you find a point where the signal disappears or otherwise becomes unacceptable, you have isolated a trouble stage. Look for a faulty component in that stage of the circuit. All earlier stages have been cleared of suspicion.

Most signal generators put out a single frequency signal of a specific waveshape (sine waves, square waves, and triangular waves are the most commonly used). Some signal generators sweep through a range of frequencies. This type is used to test the frequency response of the circuit under test.

FUNCTION GENERATORS

A function generator is basically a special form of the signal generator. Most signal generators put out just a single waveshape. A function generator, on the other hand, lets the technician choose from three or four different waveshapes. For certain tests, a square wave, for example, might be better than sine wave, or vice versa.

Function generators also tend to have a wider frequency range than simple signal generators. Many function generators are capable of putting out signals ranging from a fraction of a hertz on up to several megahertz.

RF SIGNAL GENERATORS

For radio frequency work, an rf signal generator is used. This is similar in concept to regular (audio and ultra-audio) signal generators, but the output frequency is much higher.

The output from an rf signal generator is almost always a sine wave. Usually there is the capability to modulate the main output signal with another signal in the audio range, and the technician can select between AM (amplitude modulation) or FM (frequency modulation).

Because of the high frequencies being produced, an rf signal generator must be very well shielded. Usually it must be type-approved by the FCC. Rf signal generators are usually crystal-controlled for accuracy and stability.

SIGNAL TRACERS

A signal tracer is often used in conjunction with a signal generator in testing audio equipment. It can also be used by itself. This is a device for detecting the presence of a signal. Usually it's not much more than a small amplifier and speaker with test leads for taking its input from the circuit point under test.

Often a signal tracer is used in the reverse manner than the method described under "Signal Generators" in this chapter. The signal tracer is first connected to the circuit's output (usually the speaker terminals). It is progressively moved backward through the circuit, stage by stage, until the signal has been found. If there is no signal at the output of stage 4, but the signal is fine at the output of stage 3, then the problem is in stage 4.

CAPACITANCE METERS

Until fairly recently, capacitor testers were not found on too many electronics servicing workbenches. There are several reasons for this. The early capacitor testers were bulky and expensive. More importantly, they really weren't good for very much. Most were basically "go/no-go" type testers that checked for shorts or opens, and perhaps gave a rough measurement of leakage. Some of the top units also gave a (very) approximate reading of the capacitance range.

Shorted and open capacitors can almost always be pinpointed in-circuit without a capacitor tester. Many of the tests described in the following chapters do this. It also isn't very difficult to isolate a leaky capacitor with in-circuit voltage and current tests. The early capacitor testers could be handy in certain circumstances, but for most technicians they were an unnecessary luxury. It was nice to have one, but it was no hardship to live without it.

The modern capacitance meter, however, is a completely different animal. These are essentially digital devices that display the actual capacitance numerically and with great precision. Most capacitance meters work by measuring the time constant of the capacitor being tested. The time constant is the time required for the capacitor to charge up to two-thirds of the applied voltage through a given resistance. The resistance, of course, is determined by the meter's circuitry and is a constant.

For many applications, the exact value of capacitance is not terribly important. Circuit resistances are usually much more critical. Most resistors used in circuits today have a tolerance of 5 percent, while 20 percent tolerances are typical for capacitors. However, sometimes a capacitor can change its value outside its tolerance range and throw off circuit performance.

The capacitance value does become critical in frequency determining circuits, such as filters, tuners, and oscillators. Some applications don't care so much about the actual capacitances as much as whether or not two or more capacitances in the circuit are closely matched.

A capacitance meter is also useful for hunting down stray capacitances in a circuit. Remember, a capacitor is really nothing more than two conductors separated by a dielectric (insulator). Stray capacitances can show up almost anywhere. Even air can serve as a dielectric. Sometimes stray capacitances are at fault in older equipment that has worked fine for years but suddenly starts behaving erratically. When the stray capacitance is located, it can usually be eliminated by repositioning connecting wires or replacing (or adding) shielding between the two conductors that are acting like capacitor plates. Adjacent traces on a pc board are particularly prone to stray capacitances.

Another use for a capacitance meter is to test the quality of a length of cable. Most standard types of cable (coax, twin-lead, ribbon, etc.) have characteristic capacitances per foot (or per meter). Just measure the capacitance across the cable and divide by the length. You should get the specified capacitance-per-unit-length value, or something very close to it.

$$\text{Capacitance-per-unit-length} = \frac{\text{Total Capacitance}}{\text{Length (in number of units)}}$$

Make sure that you use the same length unit for both halves of the equation. This equation can also be reversed to get an estimate of an unknown length of presumably good cable.

$$\text{Length (in number of units)} = \frac{\text{Total Capacitance}}{\text{Cap.-per-unit-length}}$$

As you can see, the modern capacitance meter is a truly versatile and useful piece of test equipment, made possible by digital technology. An accurate capacitance meter of this type would be highly impractical, if possible at all.

FREQUENCY COUNTERS

It seems the frequency counter is the status symbol for electronics technicians. Everybody wants one, even if they're not entirely sure what they want it for.

If you work primarily in general TV servicing, you probably wouldn't use a frequency counter very often. On the rare occasions when you need to measure a frequency, you can use your oscilloscope. Just count the number of cycles displayed and divide by the time base.

On the other hand, in some applications a frequency counter can be extremely valuable. Two such applications include musical instrument servicing and working with rf transmitters. In rf transmitters, precise frequencies are required by law. A frequency counter is a handy tool for making sure that oscillators, multipliers and frequency synthesizers are working properly and are tuned correctly.

All modern frequency counters are digital devices. The measured frequency is displayed directly in numerical form. The typical input impedance of a frequency counter is about one megohm, so loading is minimal. Most commercially available frequency counters measure frequencies up to about 50 MHz or 100 MHz, which should be sufficient for most servicing work. However, if you need to measure higher frequencies, accessory prescalers are available to extend the range of a frequency counter.

SERVICING TEST EQUIPMENT

Any piece of electronic equipment is likely to require servicing sooner or later. Though most people don't think about it, that obviously includes test equipment.

Conceptually, at least, servicing a piece of test equipment is no more difficult than any other type of electronic circuit. There are two special types of difficulties involved. The first is that sometimes you need the instrument being tested to perform the necessary tests. Whenever possible, it is a good idea to have spares of general purpose equipment such as multimeters. You need a second multimeter to find a defective component in your first multimeter. Besides, if you have only one multimeter and it breaks, virtually all the rest of your servicing work will come to a standstill until you get it fixed. You may have to order a part to repair the broken multimeter and that might take a while.

Multimeters are inexpensive enough that most technicians can afford two or three. The spare units need not be deluxe models with all of the special features of your primary meter. They just need to be sufficient to get you by in an emergency.

Some types of equipment are too expensive to have spares on hand. For example, unless you work in a large shop with several staff technicians, you will probably have just one oscilloscope. If it develops a fault, you will need to borrow or rent one until yours can be repaired or replaced. It is a good idea to check out the availability of an emergency replacement before you are faced with an actual emergency.

Try to make an arrangement with nearby technicians to borrow or rent their scope in an emergency, while agreeing to reciprocate the favor if necessary.

If this is not feasible, find out what rental companies handle the necessary equipment in your area. Get a copy of their rates and update it periodically.

Do this leg-work before you need it to avoid frustration and desperation in an emergency.

VOMs are probably the easiest piece of test equipment to repair, because there really isn't very much inside them. They are basically made up of switches, a handful of precision resistors, a diode or two and indicators. Common faults include broken wires leading from the switches, shorts, or burned-out diodes or resistors. Burned-out components are almost always caused by accidentally feeding too high a voltage or current to the VOM or trying to operate it on the wrong range. Trying to measure a moderate to large ac voltage while the meter is set for ohms, for example, will almost surely result in disaster. Unfortunately, this is very easy to do. Almost every experienced technician I've ever met has blown out at least one meter at some point in his career. If you haven't made this type of goof yet, don't feel too smug. The odds are good that you'll do it in a moment of carelessness sooner or later.

Unless the over-voltage or current is extremely high, this type of problem is usually fairly easy to repair. It's just a matter of locating the burned-out component(s) and replacing them. Since there aren't that many components in the entire VOM circuit, you could test every component individually in just an hour or so. You can make better use of your time, however, by using a little logic to eliminate some of the components as suspects.

In most cases, the damaged VOM will work on some ranges or functions but not on others. Obviously any component included in a working range must be good, so there is no point checking it.

Often one or two resistors or diodes will be visibly burned. Sometimes a burned component looks okay, but can be located by smell, especially right after it was damaged. But don't rely on eyeball or nose tests by themselves. Multiple components might have been damaged. A resistor might change value or be internally cracked without being visibly burned. Bad diodes usually can't be detected without a resistance test of the pn junction. A component that looks bad probably is, but a component that looks good might not be. Don't be too quick to jump to conclusions.

There is one part of the VOM that is tough to repair—the meter itself. It can be damaged by too large a signal or by a physical shock (such as being dropped). Occasionally, the only problem is a bent pointer needle. If you have a very sure hand, you might try repairing this. Open up the meter's housing and *carefully* bend the meter back into the correct position. Do not use too much force, or you'll make the problem worse. You could break the pointer needle off or bend it irreparably. The pointer needle must be absolutely straight, or the meter will be useless.

In some damaged meters, the driving coil might be burned out. This is usually not repairable, and the meter itself must be replaced. For a VOM, this usually means buying a whole new instrument, because the meter makes up the bulk of the unit's cost. Do not try substituting a cheaper meter unless you are really desperate. An unauthorized replacement probably won't fit mechanically very well, so it will be more prone to physical damage, and probably won't last very long. More importantly, its electrical characteristics probably won't be exactly the same as the original, adversely affecting the accuracy of the unit.

VTVMs or other electronic multimeters (FETVMs, etc.) aren't much more complicated to service, but the circuitry does include one or more active amplification stages. These circuits can be repaired with standard voltage/current measurements and signal-tracing techniques. For some electronic multimeters, it might be worthwhile to order a replacement meter movement from the manufacturer rather than junking the instrument and buying a new one. Use an exact replacement, or you'll just be asking for trouble.

Repairing a multimeter isn't particularly difficult in most cases. But in order to put it back into service, you must first confirm its accuracy. The unit should be recalibrated after any repair or adjustment. It is a good idea to perform calibration procedures any time the instrument's case has been opened. Occasionally, the only repair needed on an apparently defective multimeter is recalibration.

First make sure the meter's pointer swings smoothly over its entire range. To check this, set the multimeter to one of its ohms ranges. With the test leads held apart, the pointer should be all the way over to the right end of the scale (infinity). Now touch the two lead probes together to create a short. The pointer should swing smoothly to the far left (zero). Bring the leads together and apart several times while watching the pointer closely. It should swing back and forth smoothly without sticking anywhere along its path. If it does stick, the most likely cause is some debris in pivot base. Open the meter's housing and carefully remove any foreign matter. You'll need a sharp eye; any very tiny particle could cause a problem.

Now set the multimeter to a voltage setting and make certain it is reading the correct value. You need some sort of reference voltage—a battery will do, if it is fresh. A mercury cell is the best because it holds a very stable voltage, but an ordinary flashlight cell will do if it has never been used or hasn't been sitting on the shelf too long. A new flashlight battery puts out a reasonably precise 1.56 volts. Adjust the meter's calibration control (usually an internal trim pot) until the meter displays the correct reading.

An oscilloscope is best calibrated with a square-wave signal of a known level fed to the vertical amplifier inputs. Watch the display closely for any distortion, especially rounded corners. This could be caused by faulty coupling capacitors in the signal path.

Signal generators can be tested with an oscilloscope and/or a frequency counter. Make sure the output frequency correctly corresponds to the nominal frequency of the unit or the setting of the front panel control(s) on a variable unit. On an oscilloscope that is known to be correctly calibrated, check the output waveform for purity, low distortion, and symmetry.

2
Dc Voltage Tests
and Power Supplies

Dc voltage tests are usually the first step in troubleshooting, and the fact that they are one of the easiest tests adds to this reasoning. The dc voltages of a device usually provide the basis for operation. The remainder of this chapter focuses on the wide range of power supplies and their circuits.

DC VOLTAGE MEASUREMENTS

To measure dc voltage, we use a dc voltmeter. This sounds a little obvious, but there are several different types of voltmeters, and there are some circuits that need the right type if test readings are to match a standard.

There are three basic types of voltmeters in wide use today. The first is the *analog voltmeter*, which uses a swinging needle over a calibrated scale (*d'Arsonval* movement). These meters have an adjustable resistance in series with the leads to allow proper range selection.

The second major type is the *vacuum-tube voltmeter* (VTVM), or its transistorized counterpart, which uses an amplifier to drive the meter movement.

The third and final major type is a relatively recent development. The *digital voltmeter* uses an analog-to-digital (A/D) converter to directly display the measured value in numeric form on a *light-emitting diode* (LED) or *liquid crystal display* (LCD) read-out panel. No mechanical meter movement is used at all.

In the analog type, the meter movement and its series resistance are hooked directly across the circuit under test. For measurements with a VTVM, a very large voltage-divider resistor is hooked across the circuit; the meter is driven by a dc amplifier whose input is "tapped down" on this voltage divider to obtain the desired range. Digital voltmeters are generally used in the same manner as mechanical *volt-ohm-milliammeters* (VOMs). That is, the test leads are connected directly across the circuit or component to be tested. Each type of meter has its own uses, advantages, and disadvantages.

When it comes to using voltmeters, the most important difference between the VOM (analog) type and the VTVM (tube) type is in their *input impedance*, or resistance. The input impedance of the meter can sometimes affect the voltage reading you get. The 1000-ohms-per-volt meter—a typical analog voltmeter—must offer a total resistance of 1,000 ohms for every volt of the full-scale range you want to be able to read. Such a meter, set to a 10-volt (full-scale) range, would have a total resistance of only 10,000 ohms; on a 200-volt range, it would have a resistance of 200,000 ohms. The 20,000-ohms-per-volt meter—another analog voltmeter—is much more sensitive. With this meter, the 200-volt range has a total resistance of 4 million ohms (4 M).

The unofficial standard for VOMs is 20,000 ohms-per-volt. Many modern VOMs have input resistance ratings of 50,000 ohms-per-volt, or even 100,000 ohms-per-volt. The earlier 1,000-ohms-per-volt VOMs are rarely used, except in very non-critical applications.

The usual vacuum-tube voltmeter uses an input voltage divider whose total resistance is 11 megohms. Some recent instruments have up to 16 megohms input impedance. This resistance remains the same for every voltage range the meter may have.

Transistorized VTVMs have similar input resistance ratings. *Field effect transistors* (FETs) are often used in the input circuits because of their high input impedances and because they are true *voltage* amplifying devices. Ordinary bipolar transistors are current amplifiers.

The input resistance of the voltmeter may or may not be of significance, depending on the specific measurement being made.

In power supplies, batteries, and other low-impedance, high-current circuits, the meter impedance—whether high or low—will not affect the circuit under test. If the voltage to be read has an ample current reserve, the type of meter used has nothing at all to do with it. They will all read the same.

On the other hand, in very high-impedance circuits—such as the plate circuit of a tube where the plate load resistance may be up to, say, 1 megohm—meter resistance can make a lot of difference! Let's take a tube with a rated plate voltage of 50 volts, which is supplied from a 150-volt source through a 1 M resistor. Ohm's law says that this means a normal plate current of .0001 amp, or 100 *microamps* (μA).

If you try to read the plate voltage with a 1,000-ohms-per-volt meter on a 50-volt scale, total meter resistance will be 50,000 ohms. Placing this low resistance across the tube results in a shunting effect; current through the plate load resistor will increase because it now has a 50,000-ohm path from plate to ground. Since it doesn't take very much current through the 1 M resistance between the plate and B+ to give a terrific voltage drop, the indicated plate voltage will probably be close to zero!

If you use a 20,000-ohms-per-volt meter on a 50-volt scale, our meter resistance will be 1 megohm. Putting this in parallel with the tube, you'd read about three-fourths of the actual voltage, or about 37.5 volts. This is still not accurate enough, unless you consider the voltage drop *caused by the voltmeter*.

However, if we hook a vacuum-tube voltmeter with an input impedance of 16 megohms across the tube, it will be more accurate. Now that the shunt resistance through the meter is much higher than the effective resistance of the tube, it would read much closer to the actual value. In fact, since most voltage readings in such circuits are now made with VTVM's at the factory, we'd probably read 50 volts.

When testing a circuit, you might find a little box in the lower corner of the schematic diagram showing which type of meter was used in taking the standard test-voltage readings. In some of the older sets, you'll find that a 20,000-ohms-per-volt meter *was* used—in which case; if you use a VTVM, all voltages will *seem* high. So, to get an accurate voltage reading, we must always know the test conditions and the type of instrument used to make the standard readings.

Digital voltmeters are normally used in the same manner as VOMs, but in terms of their electrical characteristics they are closer to VTVMs. Digital voltmeters generally feature a very high input impedance. One megohm-per-volt is typical. A digital voltmeter makes precise readings very easy, because a numerical value is displayed directly, and there is none of the inherent ambiguity involved in determining the exact position of a meter pointer on a scale.

On the other hand, in many cases you need to monitor a continuously changing value. The trend of the change can be easily monitored by watching the up or down movement of a meter pointer, but on a digital voltmeter you would only get a blur of meaningless numbers. The best type of voltmeter to use depends on the nature of the specific application at hand.

THE FULL-WAVE POWER SUPPLY

Figure 2-1 shows a full-wave rectifier power supply with a step-up power transformer. This type of circuit is used in many TV, PA, and radio circuits.

Let's look at the ac/dc voltage relations in it. To read the supply voltage for the TV or amplifier (or the "load"), put the positive lead of the voltmeter on the filter output, which is one terminal of the choke (coil). Always set the meter to a scale that will show *more* than the voltage you expect to read. In typical TV circuits, this is between 300 and 500 volts. Start with the meter on a 0- to 500-volt dc scale.

Notice one odd thing: while putting in 300 volts ac at the transformer secondary, you get 350 volts dc at the filter input, after rectification. Are we getting something for nothing? No. This apparent voltage increase is due to a difference in the way the voltmeter reads ac and dc. Ac voltage is read as an *effective* or *rms* (root-mean-square) which is only 0.707 of the actual peak or maximum voltage in the waveform. (Conversely, the peak value is 1.414 times the rms value.) By rectifying the ac voltage, you get a series of dc pulses that reach the *peak* value of the input voltage. Here, this would be 300×1.414 or about 420 volts.

When feeding this pulsating dc voltage to the large electrolytic input filter capacitor C1, it charges the capacitor to the peak voltage, less the "rectifier drop." The exact amount of voltage dropped across the rectifiers will depend on the characteristics of the specific devices used. For convenience in this discussion, assume the rectifier drop is about 50 volts. When drawing current out of the power supply circuit to feed the load, this drops the voltage a little more, winding up with about 350 volts dc at the filter input. Drawing the load current through the filter choke gives a small voltage drop, so you read about 335 volts dc at the filter output. In some circuits, you'll find resistors used in place of the choke. The higher dc resistance makes the drop within the filter higher, so you get a lower voltage at the filter output.

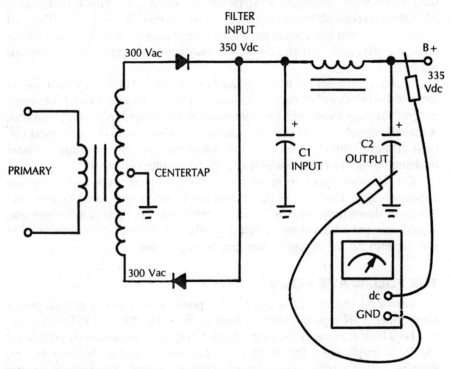

Fig. 2-1. A full-wave rectifier power supply circuit with a step-up power transformer like this is used in many TV, PA, and radio circuits.

FULL-WAVE POWER SUPPLY WITH FLOATING B–

The circuit shown in Fig. 2-2 is exactly the same as the one in Fig. 2-1, except the center tap of the power transformer secondary does not go directly to ground but through a pair of resistors. Since all of the load current flows though these resistors, a voltage drop will develop across them, and this will make B– negative with respect to *ground*.

If we take the B+ voltage measurement as in the previous section—that is, ground to B+—you get the B+ voltage of 335 volts *less* the value of the highest negative voltage. So at the filter output, you read 335 – 40, or 295 volts, to ground.

To read the *total* B+ voltage, which is the same as before, hook the voltmeter leads as shown in Fig. 1-2—negative to B– and positive to B+. However, in almost all cases take these readings to ground as a positive and a negative voltage, and add them up in your head. The voltages will be shown on the service data as so many volts positive (+) and so many volts negative (–), and this is really all we need to check.

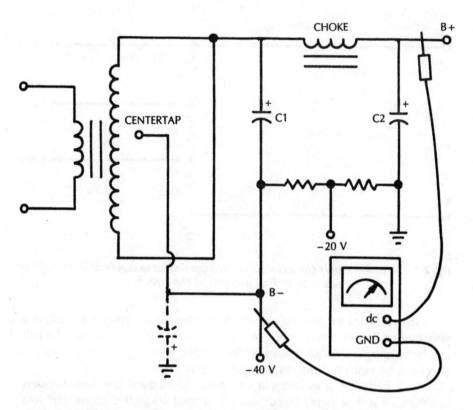

Fig. 2-2. By adding a center tap to the power transformer secondary, a full-wave power supply with floating B– is created.

There is one thing you must watch in this circuit. When replacing filter capacitors, notice that the negative terminal of the *input* filter does *not* go to ground as before, but goes to the B– point. This is necessary to get proper filtering.

POWER SUPPLIES WITH VOLTAGE DIVIDERS: 1

In actual circuits, we need to divide the B+ voltage to get the proper values on the different circuits. Therefore, you must use a circuit called a *voltage divider*. Such a circuit is shown in Fig. 2-3. With the B+ voltage from the full-wave power supply just discussed, all we need to do is make the + and − connections to the divider.

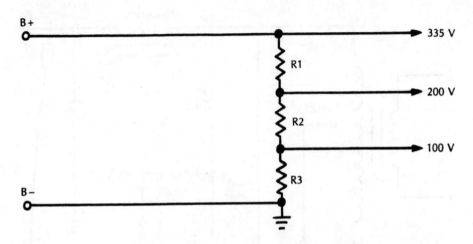

Fig. 2-3. *This power supply circuit includes a voltage divider to supply different voltage values to various portions of the circuit.*

When hooking the series of resistors across the supply, you get a small, constant current flow in the resistors. This bleeder current helps to stabilize the voltage by furnishing a constant load on the supply. You can tap off any voltage you need by tapping down on the voltage divider.

You'll find a lot of variation in resistance values used in voltage dividers, but the total will probably be somewhere around 40-50,000 ohms. The total resistance is chosen so the bleeder current will not be too high, which could overload the power supply. The values of the individual resistors are chosen to divide the total voltage as desired.

The resistors can be checked by reading the voltages at the taps and vice versa. Here, the B− is grounded, so put the negative lead of the voltmeter to ground, and read the voltages at the taps. For instance, you might get something like this: The 335-volt tap reads 345 volts; the 200-volt tap reads 210 volts; the 100-volt tap reads zero. What's the trouble? R2 is open. How do you know? Because of the characteristics of the readings; note that the taps that still read voltage have higher readings than normal. This means that part of the normal load has been lost. The bleeder current has also ceased, since the circuit from B+ to ground is open, and this raises the voltages still more. Double-check by turning the set off and checking the resistance of R2 with an ohmmeter.

Now let's see what happens if we find *almost* the same symptoms as in the preceding unit. Suppose the 335-volt tap reads 320 volts, the 200-volt tap reads 180 volts, and the 100-volt tap reads zero again. Note that these voltages are now lower than normal. You'll notice another typical symptom if you touch the resistors—R1 and R2 are now very hot, and R3 is cold.

If you turn the set off and take resistance readings, all of the resistors themselves are correct. However, when you read the resistance from the 100-volt tap to ground, there's a zero reading. There is the trouble! Check the bypass and filter capacitors in the circuit; one of these has shorted out.

The key clue, besides the overheating of the other resistors, is the low voltages. When the capacitor shorted, it took R3 out of the circuit completely. There is that much less resistance from B+ to ground. This means that the bleeder current will go up, and the total voltage will go down. In many cases, this kind of overload will be enough to burn out R1 or R2. So if you find an open resistor at any point in a voltage divider circuit, check on its *load* side (the end away from the source of voltage) to make sure that there is no short circuit there that could burn out a newly installed resistor.

POWER SUPPLIES WITH DROPPING RESISTORS

Figure 2-4 shows the equivalent of the previous circuits, but with the resistors arranged a little differently. Here we have individual resistors connected into each voltage-supply circuit. The end result is the same—each circuit gets its proper voltage. Without the multiple-tapped voltage divider resistors, there is no bleeder current in such a circuit.

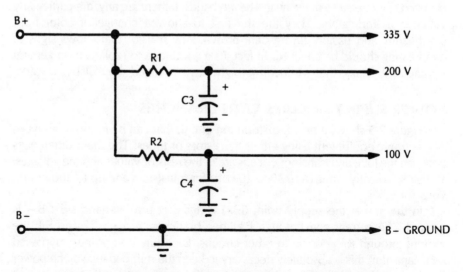

Fig. 2-4. This power supply circuit features dropping resistors.

The dropping resistors are chosen so that their values will give the correct voltage to each circuit. This feature gives us a very good clue when there's trouble. If the voltage on any tap is low or high, then check that circuit first to see why. The rest of the voltages won't be affected as much, but will be changed to some extent.

First check the supply voltage (this rule holds for all power-supply troubleshooting). Without the right supply voltage, the other voltage measurements are meaningless. Use some of the same obvious clues as before. For example, if some of the dropping resistors are very hot and the supply voltage is low, you know immediately that there's a short circuit somewhere.

If the supply voltage is low, but there are no signs of overheating, then look for something that is weak—a leaky or shorted diode, an open input filter capacitor, a weak selenium rectifier, or anything that would reduce the ability of the power supply to deliver the right amount of current.

These same tests apply to all circuits: First check the supply voltage, then the individual tap voltages. In the circuit shown, a short in one load circuit will kill the voltage at that tap, but it won't make the others drop as much as before. For example, if C3 is shorted, the 200-volt line will read zero, the 335- and 100-volt lines will go down by about 10 percent. The key clue will be R1; it will be very hot. If the short has existed long enough to cause R1 to burn out, then the 200-volt line will read zero, but the other voltages will be above normal because of the loss of the normal load current in the 200-volt line.

Many sets use this circuit. The average dropping resistor will be a 2-watt carbon type. If it's been overheated, you'll notice a decided change in its appearance. The case will be darkened, and the color-coding paint will have changed color because of the heat. The red bands, if there are any, are particularly valuable as indicators. They are the first to show a change of color from overheating and usually turn a dark brown. Any resistor that shows signs of overheating should be checked. In fact, its a good idea to replace it on general principles, because the overloading might have changed the resistance value.

POWER SUPPLY CIRCUITS WITH BRANCHES

Figure 2-5 shows a power-distribution circuit that can be found in all types of electronic equipment using either transistors or tubes. The main differences between the tube and transistor circuits are in the resistor values and the voltages. Voltages typically range from 30 to 40 volts for transistors and up to 300 or 400 volts for tubes.

In the figure, the supply point (B+) is again at the left-hand side, B− is ground. The current goes through R1, then branches through R2 and R4; the current through R3 goes on to other circuits. Each junction point is bypassed by a capacitor; this is absolutely necessary to keep the rf impedance of the power supply very low. Since this distribution circuit is common to all stages, it must prevent any interaction between the various circuits. If not, you get feedback and the equipment will oscillate.

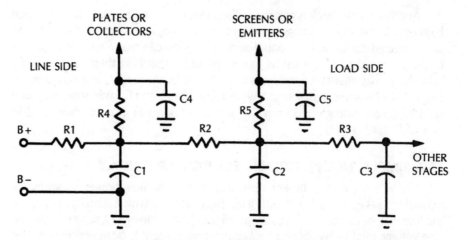

Fig. 2-5. A power-distribution circuit like this one is used in both tube and transistor circuits.

The capacitors are the most common cause of voltage troubles. If any one of them shorts, it will kill all voltages past the point where it is connected in the circuit. However, there's a quick and easy way to find a shorted capacitor.

Here B+ is the supply side or line side of the network. The right-hand side of the circuit is the load side—the stages that need the voltage. If we know where the current is coming from and what it goes through on the way, we can tell what's wrong when it doesn't get to where it should!

For instance, if C2 shorts, you'll read zero voltage at the junction of R2, R3, and R5. Clue No. 1: Voltage is present at the junction of R1, R2, and R4, but it isn't as high as it should be. Clue No. 2 is easy to spot: R2 is probably getting very hot. R3 and R5 are cold. So look on the load side of R2 for the short. The current must go through the last hot resistor in the circuit before it gets to the short. Turning the power off and taking an ohmmeter reading and look for a zero resistance to ground; such a reading is always incorrect in a voltage supply circuit, so there's the trouble. To confirm this diagnosis, disconnect C2 and check it. The short disappears from the resistor network, and C2 shows a high leakage or short.

When you're hunting for a short in the power supply, start at the power supply (B+) and go from there toward the load(s). Remember this procedure, because it's always the easiest way to find the cause of the trouble. Use the schematic diagram. Trace the power-supply circuits to each stage through the dropping resistors past any bypass capacitors that could short out until you get to the point where there's no voltage.

Transformer-powered circuits have normal resistances of about 20,000 ohms to ground. This is mainly the leakage resistance of the electrolytic filter capacitors. In silicon-rectifier or voltage-doubler circuits, you might have to disconnect the rectifiers; they offer a very low back resistance that can falsely indicate a short circuit.

Another quick check is to simply disconnect some to the loads and see what happens. If the supply voltage suddenly jumps back to normal, the last load disconnected could be the trouble spot. Also, you can make mental additions of the various dropping resistors in the circuit by checking their values on the schematic; take resistance readings at each end of the circuit and compare. In Fig. 2-5, for instance, if all resistors were 1,000 ohms, and C1 was shorted, you'd read 1,000 ohms to ground from B+ and 2,000 ohms to ground from the load end of R3 (R2 + R3).

THE SIMPLE TRANSFORMERLESS POWER SUPPLY

Many types of equipment now use transformerless power supplies for economy's sake. Instead of stepping up the ac voltage with a transformer, apply the line voltage directly to a rectifier at the standard value of 117 volts rms. Actual line voltage might vary, of course, but most equipment is designed to work, as stated on the rating plate at "105-120 volts ac."

Figure 2-6 shows the simplest possible circuit—a half-wave rectifier and filter circuit (the filter is the same as before). Here again, we find the dc voltage higher than the ac input. With a 117-volt ac input, there's about 145 volts at rectifier and about 135 volts at the filter output. The input capacitor (C1) will be a large electrolytic, usually from 60 to 100 microfarad (μF). A resistor of about 1,500 ohms is used as a choke.

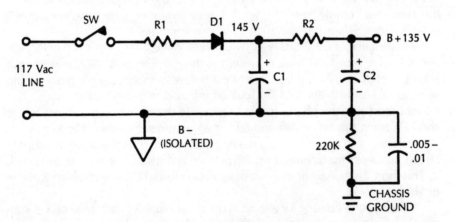

Fig. 2-6. The simplest possible transformerless power supply circuit is a half-wave rectifier and filter.

Notice one important thing about this and all "line-connected" circuits. You can't use the chassis as the B− any more; if you did, this would mean that it *could* be connected to the hot side of the ac line if the line plug was plugged in "wrong." This would be very dangerous to the user. So B− is isolated from the chassis; this B− point is shown by the hollow triangle symbol. This point is usually returned to the chassis through an RC network consisting of a large-value resistor shunted by a bypass capacitor. The chassis-ground symbol is the same as usual.

To measure dc voltages in this circuit, take the readings between B+ and B–; measurements between B+ and chassis won't be accurate because of the large series resistor. One fast way to find the negative terminal is to put the voltmeter negative prod on the negative terminal of the electrolytic capacitor. In can-type electrolytics (with one exception) the can is always negative. In cardboard-tube types, the black wire will be the negative if standard color coding is used.

THE HALF-WAVE VOLTAGE DOUBLER FOR TELEVISION

The 135 volts from a simple transformerless power supply isn't high enough for TV circuits, so in this case a circuit is needed that will give us a higher voltage. Such a circuit is the voltage doubler, shown in Fig. 2-7. There are several types of voltage-doubler circuits, but this one is the most popular, probably because it's the simplest. It is called a half-wave voltage doubler, because it uses each half of the incoming line voltage to charge a separate capacitor; the two capacitors are then discharged in such a way that the voltages add up. This is the basis of all voltage-doubling circuits.

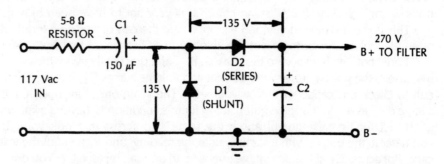

Fig. 2-7. A voltage doubler is used to obtain a higher supply voltage.

Although they tell us that an electrolytic capacitor should never be connected to the ac line, here we seem to have one hooked right to it: C1 is in series with the ac input. However, following the return circuit, you can see that the capacitor is never subjected to a true alternating voltage. On the first half-cycle of voltage, the polarity is such that D1 is conducting. Current flows into C1, then out through D1, which is the "shunt" rectifier, and back to the other side of the ac line (note that this is a unidirectional current). When we say that current flows through C1, we mean only this one-direction *charging* current. Such a current flows in all capacitors while they are charging to an applied voltage.

Now, C1 is charged to about 135 volts, and the next half-cycle of line voltage has such a polarity that D1 is not conducting, but D2 is. So, the current from the line flows on through D2 and charges C2; at the same time, C1 *discharges* through the same circuit, and its voltage is added to the charge on C2 because the polarity is the same (positive with respect to B−), C2 charges up to double the line voltage and you get about 270 volts instead of the 135 that was in the ordinary half-wave rectifier circuit.

The voltages here are read to an isolated B− point, as shown. The most important thing to look for in voltage readings here is balance. If both rectifiers and both capacitors are good, you will be able to read the dc voltages as shown: 135 volts across D1, from B− to the center tap of the two rectifiers, and an equal 135 volts directly across D2, from center tap (place the negative meter lead here) to positive or B+. If you read 135 volts across one rectifier and only about 80 volts across the other, the chances are that the "low" rectifier is faulty.

If the input capacitor loses capacitance or opens, the dc voltage will drop to less than half normal or disappear entirely. If C2 opens or loses capacitance, the dc voltage will drop quite a bit, and there will be a great increase in the hum level. The small resistor seen in the ac input is called a surge resistor and is a fusible type; although it is a wirewound resistor of about 4 to 5 watts in rating, it is designed to act as a fuse. If something shorts out in the power supply, it will open and prevent further damage. (A good check for this resistor in actual circuits is to carefully feel it after the set has been on for a minute or two. If it's warm, it's okay. If it's "stone cold," it is very apt to have been blown.)

This is the instance mentioned earlier where the can or negative terminal of an electrolytic capacitor is *not* B−. Notice from the circuit that *both* sides of C1 are "hot" with respect to the chassis, B−, and practically everything else! Because of the pulse nature of the current and voltage across C1, it is very difficult to check this capacitor with a voltmeter. The symptom of an open C1 is a large drop in B+ voltage; the quickest test for this condition is to shunt a known good electrolytic capacitor of about the same size across it. Hook your dc voltmeter to the output with test clips, note the reading, and then shunt the test capacitor across C1; if the dc voltage jumps up to normal, replace C1. You don't have to use the same size capacitor for *testing*. If you replace a capacitor, always use exactly the same size; but almost any size capacitor will do for a quick test. (If C1 is 150 μF, as many are, you can shunt, say, and 80-μF unit across it; a very definite rise in the dc voltage indicates that the original capacitor is bad.)

THE FULL-WAVE VOLTAGE DOUBLER

The full-wave doubler, illustrated in Fig. 2-8, is the original voltage-doubler circuit. It was used with special dual-diode vacuum tubes in old radios, and is now found in quite a few circuits with selenium or silicon rectifiers as shown here. To make the circuit easier to read, it's shown fed by an isolated transformer secondary winding—a circuit that you'll find in actual equipment in many cases.

In a few cases, this circuit is used without the isolation transformer—not too many, though, because the floating center tap makes it hard to handle with respect to the rest of the circuits.

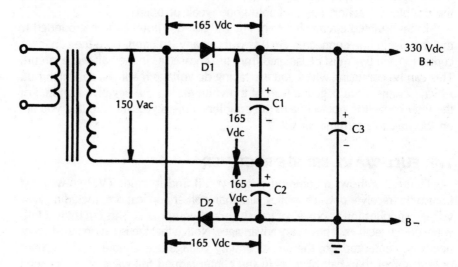

Fig. 2-8. This is a full-wave voltage divider circuit.

It works like this: On the first half-cycle of voltage from the ac supply, D1 conducts and charges C1, just as in the half-wave rectifier circuit. Because D2 is reverse biased during this time, it does nothing. On the next half-cycle, the polarity reverses; D2 conducts and charges C2, and D1 is cut off. Now two big electrolytic capacitors, each one charged to 165 volts dc, are connected in series. So, you can take the "added" voltages off and feed them to the filter as the sum, or 330 volts dc. The action of this circuit depends on the alternate-charging and series-discharging of these two capacitors. This is the basis of all voltage-doubling circuits, but it is not so apparent in the others. Here, you can see the two separate capacitors.

Tests are the same as in the half-wave circuit. Read the dc voltage across each rectifier, watching your voltmeter polarity; this reading should be 165 volts dc on each one. The same goes for capacitors C1 and C2: the voltages must be equal, because the capacitors are to be the same size. The dc output voltage is the sum of the voltages across the two capacitors. C3 is the filter capacitor.

If one of the doubler capacitors should lose capacitance or open, the circuit will not balance and the output voltage will drop quite a bit. The same thing happens if one of the rectifiers goes bad. As mentioned, the doubler capacitors must match; you'll find sizes from 40 to perhaps 200 μF used here. The size of the doubler capacitors affects the developed dc voltage. As a rule, the bigger the doubler capacitors, the higher the dc voltage, because larger capacitors will hold more charge.

The filter capacitor, C3, has nothing to do with the doubling action; similarly, the doubler capacitors have very little to do with the filtering. So if you have a bad hum with almost normal output dc voltage, the filter capacitor is weak. If you have some hum and the voltage is low (more than 25 percent), one of the doubler capacitors has probably gone weak or open.

In the isolated circuit shown in Fig. 2-8 the B− point can be grounded to the chassis. In line-connected circuits without any isolation, both the capacitor center tap and B− must be isolated from the chassis, as in the half-wave circuit. This can be confusing when you are taking dc voltage readings, so watch out. Always connect the negative lead of the voltmeter to the negative terminal of the *filter* capacitor, not to one of the doublers, unless you are *sure* that you are on the negative terminal of C2.

THE FULL-WAVE BRIDGE RECTIFIER

Figure 2-9 shows a common circuit you'll find in color TV, two-way FM transmitter-receiver power supplies, and many other applications, including low-voltage supplies for ac power of transistorized equipment. This circuit is a full-wave bridge rectifier. It has many advantages. Notice that the isolation transformer needs no center tap. So, the full secondary voltage is available at the rectifier output, rather than half of it, as in the center-tapped full-wave rectifiers seen before. This saves both size and cost in the power transformer. The circuit can be made fully isolated, and B− is usually the chassis.

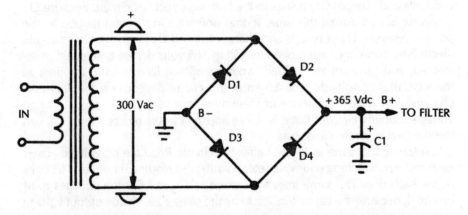

Fig. 2-9. Full-wave bridge rectifier circuits are used in many types of electronic equipment.

It works like this. On one half-cycle, one end of the secondary winding is positive, the other end negative. Current flows through D1 and D4 to the load and ground, respectively (in the same direction with respect to the load), while the other two rectifiers reverse polarity and do not conduct. On the next half-cycle, D1 and D4 reverse-bias and cut off, and current flows through D2 and D3. So, it uses both halves of the cycle, as in the full-wave circuit shown earlier.

One advantage of the bridge circuit is that there are always two rectifiers in series across the total ac voltage of the secondary. Thus, for a given output voltage, the individual rectifiers have a lower peak voltage applied to them than in other circuits. This gives them longer life expectancy, allows the use of lower ratings, etc. The ripple frequency is 120 Hz, as in other full-wave circuits.

When making replacements, look for the identification on the rectifier. Notice that there are two positives hooked together and two negatives. These are always B+ and B−, respectively. Each end of the transformer winding always gets one positive and one negative (and it doesn't matter which of these is which). In some circuits, you might find bias resistors connected between B− and chassis ground. These work in exactly the same way as in the previous full-wave circuits.

The dc output voltage here is related to the ac input voltage as it is in the full-wave tube-type rectifier circuit, except that in the bridge circuit the entire secondary voltage of the transformer is the ac input. If we have a 300-volt rms ac supply to the bridge, we will read the same proportion of increase as before; so, we get about 365 volts dc at the filter input, read between B+ and ground or B− in the circuit shown. This dc voltage is developed by charging C1, the input filter capacitor, to the peak voltage of the rectified ac output, just as in all other rectifier circuits. If the output voltage is low, this capacitor should be checked first. Once again, hook the dc voltmeter to B+, and then shunt a good electrolytic capacitor across C1. If C1 is low in capacitance, you'll see the voltage jump back up toward normal. If C1 is completely open, your dc voltage will probably read about 40 to 50 percent of normal, or even less, depending on the load current being drawn.

In solid-state circuits, a special bridge rectifier is often used. The four diodes making up the bridge are packaged in a single unit. Such devices function exactly the same as separate diodes.

VOLTAGE REGULATORS—TUBE TYPES

Frequently, dc voltages need to stay constant: to stabilize critical circuits, hold gain constant, serve as reference voltages, and so on. Regulators provide this stabilization. In vacuum-tube circuits, special voltage-regulator tubes are used. These are gas-filled; by the use of different electrode spacing, types of gas, etc., they can be made to have different output voltages. The common values are 90, 105, and 150 volts.

A typical tube-type voltage regulator circuit is shown in Fig. 2-10. A series dropping resistor is connected to the B+ supply voltage, and the tube is connected from this point directly to ground. If the supply voltage rises, the tube conducts more current; this increases the voltage drop across the series resistor, so the voltage drop across the tube stays almost the same as before. If the voltage falls too low, of course, the tube has no control (it usually goes out—that is, stops glowing).

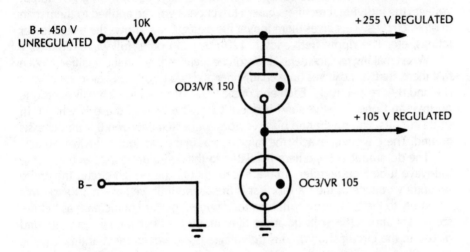

Fig. 2-10. This is a typical tube-type voltage regulator circuit.

In this circuit, taken from an oscilloscope, a type 0D3 (old designation, VR-150) and 0C3 (VR-105) have been hooked in series to give a total of 255 volts dc. A 105-volt tap comes off the 0C3 for another circuit.

A voltage-regulator tube can be used as an indicator of the presence of B+ voltage. Just look to see whether it is glowing. Some glow blue, others have an orange tint, depending on the gas used. If the regulator tube is glowing, B+ is definitely there and probably okay. Lack of glow means that the tube is defective or that there is no B+ voltage (usually the latter).

If there's a short in one of the circuits fed by a regulator tube, the overload might drop the voltage below the level at which the tube is supposed to regulate—105 volts in the case of the 105-volt 0C3—and the tube will go out. If the dropping resistor increases in value, or if the supply voltage drops below the value needed to fire the regulator tube initially (133 volts dc, minimum, for the 0C3), there will be no glow.

Voltage-regulator tubes very seldom, if ever, short out as others do because of their construction. However, if you suspect that one has, simply pull the tube and recheck voltages.

ELECTROLYTIC CAPACITORS
IN SERIES FOR HIGHER WORKING VOLTAGE

Figure 2-11 shows a circuit that you will find now and then in high-powered amplifiers and in a few transmitters. The figure shows the basic schematic. Notice that the filter circuit has a choke input for better voltage regulation. This cuts down on the peak voltage but helps to hold the output voltage much steadier.

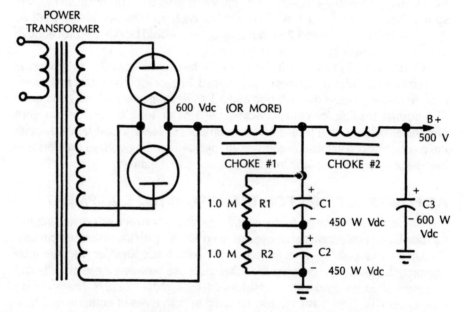

Fig. 2-11. This circuit is used in some high-power amplifiers.

In this circuit, we're getting into B + voltages that are above the normal working voltage of electrolytic capacitors. The average TV filter capacitor is rated at about 450 working volts, with a 550-volt peak surge rating. This means that the capacitor should withstand 450 volts—at turn on, for example—without blowing out. This peak rating shouldn't be exceeded or even reached for more than about 15 to 20 seconds. The working-voltage rating, for the best results and longest life, shouldn't even be met. A 450-volt capacitor, for example, shouldn't be used in circuits with steady or normal voltages of more than about 400 volts, and this is cutting it close. A 450-volt capacitor working at about 350 volts is about normal in the average well-built TV set or amplifier.

However, there are times when you must have an electrolytic that will hold a higher voltage, so put a couple of capacitors in series. This works just as with paper capacitors; you wind up with half the capacitance and twice the voltage rating.

We see this scheme used in the figure. Here, C1 and C2 are 450-volt capacitors (900 volts total) and big enough so that the resultant capacitance is enough to give adequate filtering. In a typical commercial circuit, a pair of 100-μF capacitors is used, giving 50 μF at 900 volts. One thing must be remembered whenever such units are replaced: for best results, the two capacitors must be of *exactly the same value*. If you use 100-μF and a 60-μF in series, the smaller capacitor will assume a greater percentage of the total voltage and probably blow! If you have to replace one capacitor of a pair like this, always use an exact duplicate. Actually, if one shorts out, I replace *both*, because the remaining capacitor has been severely overloaded and could fail in service soon.

The 1 M resistors hooked across the capacitors are necessary to equalize the voltages. They take no perceptible current from the circuit because they are so big. They only draw 1 mA of current for each *1,000 volts* across 1 M! So if we had 600 volts here and 2 M, our total current would be 0.3 mA (not enough to make the power transformer even run hot!).

Quick-check in this circuit: Read the dc resistance from B+ to ground at the rectifier cathodes or filaments. If you read 1 M, one of the filters is shorted. (You're reading a short through one and a 1 M resistance across the good one.) The normal reading, of course, would be 2 M or less, depending on your ohmmeter polarity. You'll read a certain amount of leakage resistance through the electrolytics. You must disconnect all loads—including voltage dividers and bleeder resistors, if any—for this test to have any meaning.

AC POWER SUPPLIES FOR TRANSISTORIZED EQUIPMENT

Essentially, there are no fundamental differences between ac power supplies for tube-based equipment and those designed for use with transistorized circuits.

Up to this point, we've talked about power supplies for vacuum-tube equipment. Now, let's see what the differences are between these circuits and ac power supplies used with transistorized equipment. Ready? *There are no differences.* The same basic circuits are used in both types of equipment. There *are* differences in the values of voltage and current, however. Tube supplies use high voltages and low currents; transistor supplies use low voltages and high currents. If a TV supply has 350 volts output at 250 mA, it draws 87.5 watts. A transistor amplifier power supply with a −25-volt output at 3.5 amps also draws 87.5 watts. There is no difference in the *power* supplied; only the values of voltage and current are different. As long as their product stays the same, the output power is the same.

In one respect, we need different test equipment for transistor circuits. In tube circuits we measure high voltages; in transistor circuits we often need to measure very small voltages—frequently a tenth of a volt or so. With tubes, voltage *tolerances* can be very high. If the rated plate voltage of a tube is 100 volts, in many cases it can vary from 90 to 110 volts without making a lot of difference in the output of the tube. (In some circuits, tolerances are as high as 20 or even 25 percent.) Hence, our dc voltmeter can have a fairly large error before it makes any difference in a tube circuit.

Transistor voltages are more critical for two reasons. First, the levels are only in the range of tenths of a volt; and second, tolerances are much tighter than with tubes. In the bias voltage reading of a typical transistor amplifier, you might find such values as these: base, 0.4 volt; emitter, 0.6 volt—giving a 0.2-volt difference. Trying to read that on a 5-volt range (the lowest on many VOM's) is demanding a great deal from an instrument whose accuracy at *full scale* might be no better than 5 percent. In the lowest quarter of the scale, it could be considerably poorer. A good VOM for this kind of work should have a 2.5-volt (or lower) low range with at least 20,000-ohms-per-volt sensitivity and 2 percent full-scale accuracy.

Changes of a fraction of 1 volt can come from a variation in the supply voltage. We find voltage-regulated power supplies quite often in better transistor equipment. Like tube circuits, the cheaper transistor units simply use "straight" power supplies and depend on the regulation of the power line to keep them within limits. However, even the best circumstances, the ac line voltage varies constantly as its load changes. The use of some kind of voltage regulation makes transistor equipment work much better. Batteries, of course, furnish a steady source of power. In high-quality equipment, such as laboratory-type transistorized voltmeters, even the *battery* supply goes through a voltage regulator!

Transistors are far more sensitive to hum than tubes. Transistors operate on very small voltage changes—which cause very large current changes for a given power output in watts. Since hum is always a voltage phenomenon, we really have to lean over backward to make sure we have taken out every last bit of hum and ripple in the power supplies. Some highly specialized circuits are in the next few pages that have done this. Voltage regulation serves a dual purpose here; since hum is voltage variation, you automatically take out the hum if you use a regulator that holds the voltage absolutely constant.

POWER SUPPLY FROM A TRANSFORMER WINDING ON A PHONO MOTOR

A simple power supply circuit found in a number of small phonographs with transistor amplifiers is shown in Fig. 2-12.

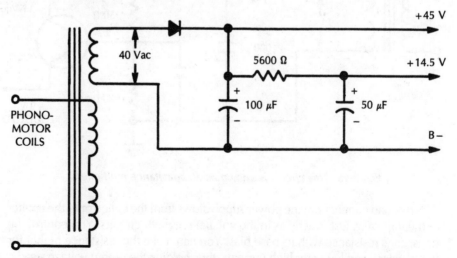

Fig. 2-12. In some small phonographs, the motor is used as part of the power supply circuit.

The novelty is in the use of a separate winding on the phonograph motor, making the motor serve as a power transformer as well. Transistors need low voltages but high currents. So using a power transformer to obtain a suitable supply voltage is much better than the only other method: a dropping resistor in series with the ac line. With a dropping resistor, heavy current changes that occur in transistor circuits would cause the supply voltage to change as well, upsetting the circuits.

In this circuit, a simple half-wave rectifier is used. In others you might find full-wave rectifiers, bridge rectifiers, and so on. They all work on the same principle as the power supplies discussed earlier.

In a variation of this circuit, used with small one-tube phonograph amplifiers, the filament of the tube is connected in series with the phono motor. The motor, of course, must be specially wound to work on the lower voltage. A typical example of such a circuit uses a 25-volt tube in series with a 92-volt motor; 92 + 25 = 117 volts ac, so the combination can be connected directly across the line. You can always tell by its behavior when this circuit is used; if the tube blows out, the motor stops!

A TRANSISTORIZED CAPACITANCE MULTIPLIER

Transistor circuits need very good power supply filtering. They also need very good voltage regulation, which is another way of saying the same thing. Very large capacitors help stabilize supply voltages; their charge furnishes extra current when sudden peaks occur in the load current, holding the voltage steady. Even with the large capacitances used in transistor circuits, you can always use more. Figure 2-13 shows a voltage regulator circuit called a capacitance multiplier.

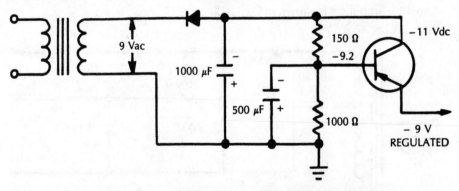

Fig. 2-13. This circuit is known as a capacitance multiplier.

The load current from the power supply flows from the collector to the emitter of the transistor, just exactly as in the voltage-regulator circuits. You control the transistor's resistance with its base bias. You can make the resistance higher for low currents, and lower for high currents, thus holding the output voltage steady.

Notice the voltage divider connected between the −11-volt supply and ground. The transistor base is hooked to the junction of the two resistors. Under normal load, the base will have a certain voltage (−9.2 volts here) that will hold the emitter voltage constant at −9 volts. It does this by controlling collector-emitter current or, looking at it another way, by controlling the resistance of the transistor.

If the load current goes up, the bias changes in response. The higher current loads the power supply; the supply voltage drops. The voltage across the voltage divider drops, and the base bias drops with it. This change makes more current flow through the transistor (lowers the resistance); so, there is now a lower voltage drop across our variable resistor and the output voltage is held where it was. (This action takes a while to describe, but it occurs in a split second.)

In effect, the circuit uses the 2-volt drop across the transistor as a reserve. When the supply voltage tends to go down, the circuit adds part of this reserve to the output voltage. The reverse is also true; if the supply voltage tries to go up, the transistor takes on a larger surplus.

Why call this circuit a capacitance multiplier? Because of its effect. If we had a very large capacitor across the power-supply output, it would hold a quantity of electrical energy; from this reservoir it could release energy whenever needed to hold the output voltage constant. The capacitance-multiplier does the same thing with a smaller capacitor. The capacitance multiplication of a given transistor is related to its current gain: the higher the gain, the better the circuit will work. (Current gain is the ratio of a small change in base bias to the corresponding change in collector-emitter current.)

The base capacitor (500 μF in the circuit shown) plays an important role. By holding a charge, it holds the base voltage as steady as possible. When the supply voltage drops, the base voltage would also drop if it weren't for this capacitor. It discharges part of its energy through the 1,000-ohm resistor, and the resulting voltage drop opposes the supply voltage drop and holds the base at the normal voltage.

CHECKING CAPACITANCE MULTIPLIER CIRCUITS

To check a capacitance-multiplier operation, check all of the dc voltages and compare them with the values given on the schematic diagram. They should be very close. Be sure to check the ac input voltage to see that it is normal; most circuits are rated at 117 volts ac input. Vary the loading. In an audio amplifier, for example, turn the volume full up for maximum load and then all the way off. This variation should have no effect on the output voltage of the capacitance-multiplier circuit.

If the output voltage is low, disconnect the load. If the voltage jumps back up to above normal, there could be a heavy overload of current being drawn by the amplifier—more than the regular circuit can handle. The load current can be measured by putting a milliammeter in series with it at the voltage-regulator output. The proper value will be given on the schematic.

Check the dc voltage between collector and emitter of the regulator transistor. If this difference is very low, or zero, check the base voltage before you decide that the regulator transistor is shorted. If, in the circuit shown, the base should be down to about −8 volts instead of the normal −9.2 volts, the transistor would be conducting as hard as it could in an effort to hold the voltage up. A wrong base voltage could result from changes in the values of the voltage-divider resistors, excessive leakage in the base electrolytic capacitor, etc.

As usual, there is an alternate trouble that can show similar symptoms. Check the control transistor for shorts and leakage. If you're not certain, take it out an check it with an ohmmeter; all transistors should show the diode effect on each pair of elements. Also, check the schematic to see if there is a shunt resistor connected across the control transistor, as shown by the dotted lines in Fig. 2-14. Such a resistor is used in some circuits to reduce the current loading on the transistor junction. It is a fairly low value resistor, and of course would upset any in-circuit tests for back resistance between collector and emitter leads.

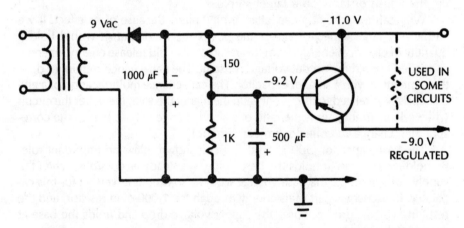

Fig. 2-14. In some cases, a shunt resistor (shown in dotted lines) is used.

If the output voltage is low—say, by about half the normal voltage—check the input filter capacitor. If this capacitor is open, there will be no reservoir effect, and the output voltage will drop drastically. It is not a good idea to bridge capacitors for test in transistor circuits; the charging pulse can cause transients that might damage transistors. For safety, turn the set off, clip the test capacitor across the suspected unit, and then turn the set on again. With the instant warmup of all transistor circuits, you won't lose any time.

If you suspect the rectifier of being weak, use the same procedure. You can clip another rectifier across the original unit—watch the polarity—and turn the set back on. If this brings the voltage back up, replace the old rectifier. Silicon rectifiers will not get weak as seleniums do; they usually short out completely when they do fail. You might find one that is completely open, but not often.

TRANSISTORIZED SERIES VOLTAGE REGULATORS

Although the circuit illustrated in Fig. 2-15 is called a transistorized series voltage regulator, you're right if you think it bears a strong resemblance to the capacitance multiplier we have just discussed. The only real difference is in the polarity of the voltages. An npn transistor is used here, since the output voltage is positive with respect to ground. The basic action is the same, though, with a few refinements. The collector-emitter "resistance" of the transistor is controlled by the base bias to keep the output voltage at the same level under varying load conditions.

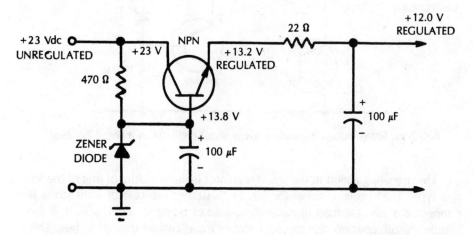

Fig. 2-15. A series voltage regulator is quite similar to the capacitance multiplier.

A zener diode is used to hold the base voltage constant. A zener diode acts as a normal diode when reverse biased until a certain voltage is reached. Then it breaks down and carries a very heavy current. This breakdown is sometimes called a *controlled avalanche.*

When this conduction occurs, the voltage drop across the diode remains essentially the same over a wide range of current flow. (A protective resistor in series is used to hold the maximum current within safe limits and prevent complete burnout of the diode.) This variable-current-constant-voltage action is the same as that of gas-filled voltage-regulator tubes.

Zener diodes come in many different voltage ratings, which are determined by their construction. To establish a certain reference voltage level, simply choose a zener that has this voltage rating. In the circuit shown, as you can see by the base voltage, the diode regulates at 13.8 volts. A 100-μF electrolytic filter capacitor helps the zener out a little.

The 22-ohm resistor and 100-μF electrolytic at the right of the figure are not a part of the regulator circuit; the resistor drops the 13.2-volt regulator output to exactly 12 volts for use by one of the circuits. The electrolytic capacitor is a filter and a bypass capacitor as well, to keep signals from mixing in the power supply and causing feedback or oscillation.

TRANSISTORIZED SHUNT VOLTAGE REGULATORS

There is one more transistorized regulator that we should mention. You won't find it used often because its efficiency isn't as good as that of the series-type regulator. The circuit, called a shunt regulator, is shown in Fig. 2-16. It is simple, but it tends to waste power.

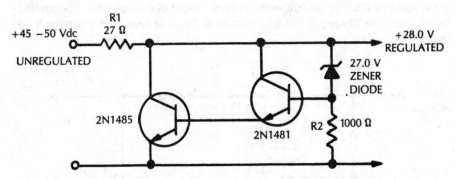

Fig. 2-16. Some voltage regulators use a shunt rather than a series hook-up.

The operating action of the shunt regulator is the opposite of that of the series type. The regulator transistor (or, in this case, the pair of transistors) is connected across the load in parallel, instead of being in series with it. If the supply voltage goes up, the regulator draws more current instead of less. This extra current increases the voltage drop across R1 and brings the voltage back down. The power consumed by the shunt transistor(s) and R1, then, can be said to be wasted.

The operating principles of series and shunt regulators are similar. The zener diode clamps the base voltage of the transistor. The voltage drop across series resistor R2 is used to signal the regulator when more of less current is needed.

A THREE-TRANSISTOR VOLTAGE REGULATOR FOR TRANSISTOR TV SETS

Transistor TV sets need very close regulation. The three-transistor circuit shown in Fig. 2-17 is used by at least one leading manufacturer. This regulator uses exactly the same principle as the single transistor circuit presented earlier, but this one is more elaborate. A pair of direct-coupled transistors is used as the regulator, controlled by an error amplifier. A 47-ohm resistor is shunted across the voltage-regulator transistor from the collector to emitter to carry part of the load.

The error amplifier makes this circuit quite sensitive to small load-voltage variations. The base is connected to the slider of the voltage-adjust control, which is a part of the voltage divider across the output. In the emitter circuit, a 12-volt zener diode clamps the voltage at this value and also provides a 12-volt regulated source for some of the TV circuits. The collector is supplied from the 75-volt line through the three resistors; it is also connected to the base of the lower transistor of the regulator. So, the collector voltage of the error amplifier sets the base voltage of the regulators.

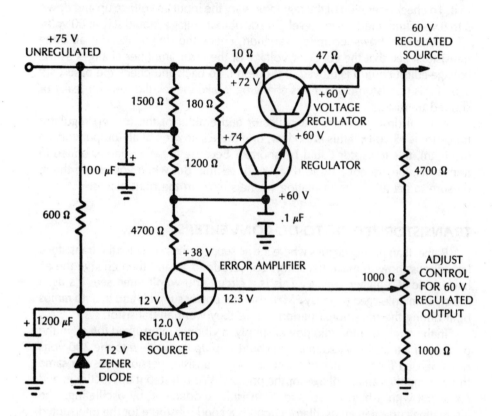

Fig. 2-17. This is a somewhat more elaborate voltage regulator circuit.

If the load current goes up, the output voltage goes down—that is, more negative. The base of the error amplifier also goes more negative. This transistor is an npn unit, so the shift in base voltage changes the bias so that *less* current flows in the collector-emitter circuit. This circuit includes the three resistors from the 75-volt unregulated source. With a lower current flowing through these resistors, the collector voltage goes more positive (it rises). This makes the base of the lower control transistor more positive. It's another npn, so the positive-going base voltage makes it draw *more* collector current. This reaction in the lower transistor is transferred and amplified by the upper one; its collector-emitter current increases (or its resistance decreases) and more current is fed to the load circuits, making the voltage come back up again. If the opposite happens and the load decreases, making the output voltage go up, just repeat all of this but reverse the polarity of each reaction.

This regulator is adjustable. Using a Variac or adjustable transformer, set the ac input at exactly 120 volts, and connect an accurate dc voltmeter across the 60-volt regulated output. With the TV set tuned to a station, adjust the voltage control to make the dc meter read exactly 60 volts, and that's all there is to it. To check operation of the regulator, vary the input ac voltage up and down 5 to 8 volts from the 120-volt level. The dc output voltage should stay at 60 volts.

If this test shows too much variation, check the dc voltages on all three transistors. See that the collector voltage of the error amplifier changes as the voltage-adjust control is moved. If it doesn't, go back and check the base voltage of this transistor to see if it is changing. Make the regular tests for leaky or shorted transistors.

Normal drop between the collector and emitter of the (upper) regulator transistor is 15 volts; if this is too low, or if there's no drop at all (output voltage too high), this transistor could be shorted. Because these are large, bolted-in transistors, they're not hard to remove for testing. But when you return them, be sure to get all of the insulating washers, etc., in the right places.

TRANSISTORIZED DC-TO-DC CONVERTERS

Rather than having circuits where the ac supplies the power for the transistors, let's discuss how to make transistors convert dc to ac and then change the ac back to dc at a higher voltage. This is a circuit you won't often see. It's used mostly in middle-aged two-way FM transmitter power supplies and in auto radios built during the transitional period before they went all-transistor.

In the original auto-radio power supply, a vibrator interrupted the dc in the primary of a transformer so that you could step up the voltage to the 200 volts or so needed for the plates of the tubes. The transistor version does the same thing; it supplies an ac voltage for the primary. We can step it up, rectify it, and come out with a high dc voltage. Transistors produce ac by oscillating. The alternating current in an oscillator circuit is a good substitute for the interrupted dc from a vibrator (it's also quieter).

These circuits are usually called dc-to-dc converters when the output is a high-voltage dc. However, there is another popular application; leave off the rectifier and filters and you have ac out. By making the output 117 volts at 60 Hz, you can use ac equipment in cars, etc. Such dc-to-ac circuits are usually called inverters or dc/ac converters.

Figure 2-18 shows a typical dc-to-dc circuit with a common collector connection. This could be a transmitter power supply, a supply for a tube-type PA system, etc. The same principle is used in all such circuits. A transistor or pair of transistors is connected to the primary of the transformer. Feedback windings make the transistors oscillate. (This is a blocking oscillator circuit, but other types can be used.)

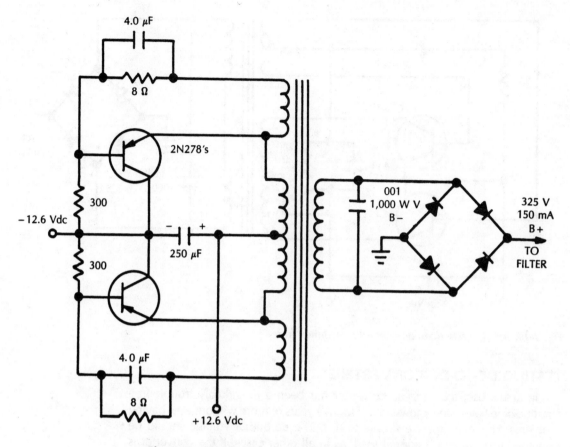

Fig. 2-18. Here is a typical dc-to-dc converter circuit.

These oscillators are often run at frequencies above 60 Hz. The higher the frequency, the smaller the transformer can be. Power-line transformers must work on 60 Hz ac; vibrators run at about 115 Hz. We can make transistor circuits run wherever we want, though. Motorola, in some of its two-way radio power supplies, uses a frequency of about 400 Hz, and others have used frequencies up into the thousands of Hz. With high-operating frequency, filter capacitors can also be much smaller for the same filtering efficiency. These characteristics make this circuit ideal for use in mobile and airborne equipment.

The output of the oscillator is a square wave. This is highly efficient and fairly easy to filter into dc. Bridge rectifiers are commonly used with standard pi-type filter circuits. Another dc-to-dc converter circuit is shown in Fig. 2-19.

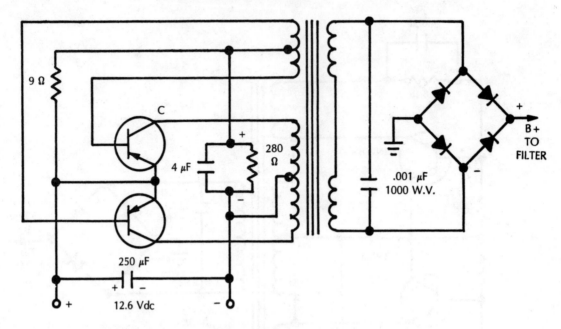

Fig. 2-19. An alternate dc-to-dc converter is shown here.

TESTING DC-TO-DC CONVERTERS

In actual use, the dc-to-dc converter has been a remarkably trouble-free circuit. Some have been known to run for five years or more without ever having problems. However, they're simple to test. The dc output voltage should be measured first under full normal load, as in all other cases. If the converter is used in a radio transmitter, the voltage should be measured with the transmitter keyed on and tuned for full rf output.

If the dc is low, check the dc input voltage first. In mobile units, make this check in the vehicle with the engine running. Many cases of false trouble have been traced to simple battery drop-off after a long test under power. If the engine is running at a fast idle, the charging system will keep the voltage up to normal as when the vehicle is in actual service. Incidentally, on a 12-volt system, this voltage should read almost 14 volts. Running the engine can make a big difference in the rf power output of a transmitter. It has been known to bring rf power from 35 watts to the full-rated 50-watt output.

The ac voltage output from the transformer can be measured if you suspect the rectifiers of being weak. However, since this is a square wave, your rectifier-type ac voltmeter will not read the true value because it's calibrated on a sine wave. If it reads about 300 volts ac, that's probably good enough. If so, then the trouble must be rectifiers or filters because the transistors are definitely in oscillation.

If the transistors are not oscillating, check all parts (since there are only five or six, it won't take long).

If either of the 4 μF capacitors in the feedback circuits is open, the oscillator will not work at all, or not at the right frequency. If it is trying to run, there will be some ac in the secondary, but not at the right frequency, so the output will be very low.

The output waveform should be in a pretty good square wave, and it *must* be balanced. If it shows a decided imbalance, look out: This means that one of the transistors is not conducting as heavily as it should. Possible causes are leaky capacitors, leakage in the transistor itself, or an off-value resistor.

The circuits shown use pnp transistors. Npn's will do the same job, but the voltages will be reversed in polarity.

IC VOLTAGE REGULATORS

More and more pieces of modern electronics equipment are using an IC voltage regulator in their power supplies. This is especially true in any equipment with digital circuitry, which requires very precise supply voltages and is very sensitive to power-line spikes.

The internal circuitry of voltage-regulator ICs is not dissimilar to the transistor voltage regulators discussed earlier in this chapter.

Most voltage regulator ICs are three-pin devices which resemble somewhat oversized transistors, as shown in Fig. 2-20. The unregulated input voltage is fed in to pin 1, and the regulated output voltage is taken off from pin 3. The middle pin is common to both the input and output circuits. In most cases, the common pin will be connected to ground, although there are some exceptions. The pin order varies from device to device, so be sure to check the manufacturer's data sheet or the schematic.

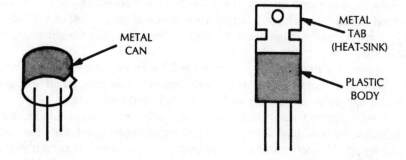

Fig. 2-20. Most voltage regulator ICs are three-pin devices that resemble over-sized transistors.

If you suspect trouble with the voltage regulator, disconnect its output from the circuit being powered. Now measure the regulator's output voltage. If this voltage is not correct, the odds are strong that the voltage regulator is defective and should be replaced. To be positive, disconnect the unregulated input line from the regulator and measure this input voltage. It should be somewhat above the desired output voltage, but there might be considerable variation here depending on the specific design. As long as this input voltage is present and "in the ballpark," it is probably good and the voltage regulator is indeed the culprit.

To test how well the unit is regulating the voltage, plug the equipment under test into a Variac. Vary the line voltage from about 90 volts to 125 volts ac. This covers the normal range of typical line voltage variation the voltage regulator should be expected to cope with. While varying the Variac's output, monitor the voltage at the regulator output. This voltage should remain constant (or fluctuate over a very small range). Voltage regulator problems can cause a variety of symptoms, such as insufficient or excessive gain or parasitic oscillations.

HOW TO READ BATTERY VOLTAGES CORRECTLY

Until now, this section has dealt with reading dc voltages in ac-powered supplies. However, don't forget the original source of power—batteries. A tremendous amount of battery-powered electronic equipment is in use today, so know how to check batteries correctly.

There's only one accurate way to read the voltage of a battery—under full load. This is easy; just turn the equipment on before taking a reading! Practically all batteries recover some voltage when the load is taken off. So if you want to know what voltage is present under actual operating conditions, check it with the equipment on and the batteries under load.

Even a dead dry-cell battery will read almost full normal voltage if you check it with a dc VTVM, which places virtually no load on the battery at all. When current is drawn from such a battery, the voltage drops to practically nothing. A brand-new standard dry cell reads about 1.64 volts. This drops to about 1.4 volts after an hour or so of use and gradually drops lower as the active materials of the battery are used up. When a cell reaches a load voltage of 1.1, it is considered "dead." Early battery radios were designed for a cutoff of 1.1 volts per cell; transistor radios might work a little past this point, but the volume will be fairly low.

Incidentally, dry batteries are figured at 1.5 volts per cell; so, a 9-volt battery would be made up of 6 cells at 1.5 volts apiece. Therefore, cutoff voltage for a 9-volt battery would be 6.6 volts under full load and so on.

Special battery testers can be obtained. Such a tester is nothing more than a dc voltmeter combined with a shunt resistor to draw current from the battery being tested. A battery tester might be useful at times, but it is certainly simple enough to just turn the equipment on and read the battery voltage with an ordinary voltmeter.

In auto radios and other equipment used in cars, the power comes from the car's storage battery. Older cars have 6-volt systems, and most new cars have 12-volt systems. The 6-volt type actually reads 6.3 volts if the battery is fully charged, and the 12-volt system, 12.6 volts.

If the engine is running, the generator or alternator should be feeding current into the battery to keep it charged; the system voltage will therefore go up. The upper voltage limit is controlled by the car's voltage regulator. With the engine running at a fast idle, the voltage shouldn't go above about 14 volts or so. If the voltage regulator isn't set properly,the voltage can go higher than this and cause damage to transistors, especially the high-power output types used in car radios now. If the system checks higher than 14 volts with the engine running, have the voltage regulator adjusted by a competent mechanic.

3

Current Tests

Current measurements aren't as simple to make as voltage measurements. The meter must become part of the circuit instead of touching across it, and this takes more time. Also, you have to be very careful—picking the wrong current range or shorting the load to ground accidentally can blow a meter very quickly. For this reason, many technicians are reluctant to use current tests. However, current measurements can give a great deal of information about a circuit in a short time. With reasonable care, there is no more risk of meter damage in this test than in any other kind of test.

Make the job of getting the meter into the circuit much easier with some special equipment. Test adapters of various kinds allow current measurement by plug-in testing without unsoldering wires. Some types of adapters can be bought ready-built, but you might have to make a few. They're well worth the little time it takes in terms of bench time saved.

Reading the input current drawn by any electrical apparatus can tell you the total wattage being consumed; simply multiply the current by the applied voltage. The actual wattage consumed is a valuable piece of information, because the rated wattage is almost always given in the service data for the apparatus. If a device is taking more power than it should, there is definitely something wrong (a leakage, a short circuit in the power supply, etc.). By measuring input current you can, for example, check a power transformer for an internal short.

The input current test works on ac and dc equipment, all the way from a tiny transistor radio to a 5,000-watt transmitter. Current measurements within a circuit are essential in transmitter testing and in all kinds of high-power work such as PA systems, high-power amplifiers, etc. In high-power transistor amplifiers, a current measurement can tell you if transistors are leaky, indicate whether bias voltages are correct, and give you other useful information.

Power Measurements

If a TV set, amplifier, etc., becomes too hot as it plays, we need to know is it actually drawing too much power, and if so, why? The input wattage measurement of any piece of ac-powered equipment is a very good clue to what's going on. A check of the actual power being consumed can be used to pin down a short or leakage in the power supply circuits, or even an open circuit—one that isn't drawing as much power as it should.

The watt is a unit of power which is always a *product*—volts multiplied by amps. So an ordinary voltmeter or ammeter will not read power—you have to take simultaneous voltage and current readings and then do the arithmetic. Of course, there are wattmeters that will do this for us. They're actually combination volt-ammeters, having a current coil in series with the line and a voltage coil across it. Both coils affect the position of the meter needle: the amount of deflection depends on the product of the currents in the two coils, so the scale can be calibrated directly in watts (see Fig. 3-1).

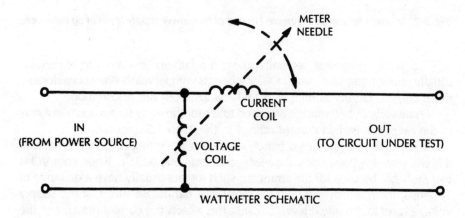

Fig. 3-1. The current through the two coils affects the amount of pointer deflection.

There are many uses for this instrument. However, wattmeters are not commonly found in service shops, because they are fairly expensive. Some shortcut tests can give the same information but use more common test equipment. Any of the wattage tests discussed can be made accurately with the three substitute testers described in a later unit.

MEASURING THE DC DRAIN OF AN AUTO RADIO

Measuring the current drain of an auto radio is probably the simplest current test. All you need is a 0-to-10-amp dc ammeter, connected to one of the power-supply leads to the auto radio, as shown in Fig. 3-2. The rated current drain for the specific set should be in the service data.

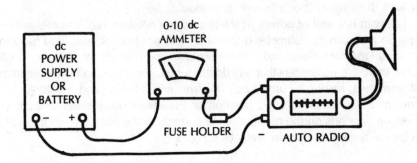

Fig. 3-2. An ammeter can be connected to one of the power supply leads of an auto radio.

Transistor equipment generally draws a relatively low amount of current, usually under 1 amp at 12 volts. Older tube sets with vibrators drew considerably more current. Drains as high as 8 to 12 amps were not uncommon.

Some early hybrid models that used tubes for the early stages and transistors in the output stage had current ratings in the 1- to 1.5-amp range.

If you use an ac-powered bench supply for operating auto radios for test, this will probably have a dc voltmeter and an ammeter built in. If not, your VOM can probably be used for the ammeter; such meters usually have a dc range of 10 amps or so, which is ample for most auto radios. Be sure that the supply *voltage* is set to the stated level, because this affects the current drawn and the wattage.

If the tone of an auto radio is not as good as it should be and the current is either more or less than the rated value, check the bias on the output transistor(s), because the output stage causes the heaviest current drain of the whole set.

DC CURRENT MEASUREMENTS
IN TRANSISTOR PORTABLE RADIOS

Current measurements are invaluable in servicing small transistor radios, particularly the subminiature types. Because of their small size, it's hard to get into these circuits. So, we take current readings, which we can do from the outside, and get all the information we can before we start taking things apart.

Practically all of the bench power supplies used for testing these radios have a dc milliammeter as well as a dc voltmeter. If yours doesn't have one, you can always use the 0-to-25- or 0-to-50-mA range of the VOM, as in Fig. 3-3.

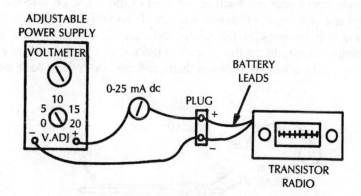

Fig. 3-3. A VOM can be used to monitor the current supplied by a bench power supply.

A set of connector harnesses can speed up the work. Take the battery terminals from a dead battery and solder a set of test leads to them, watching polarity. Now, when working on a radio using this size battery, you simply snap on the connector. It's handy to have a harness for each of the common sizes of batteries.

A set of test leads with alligator clips is very useful. For example, you can use them to connect a milliammeter in series with the radio battery by turning the battery plug sidewise and clipping on as in Fig. 3-4.

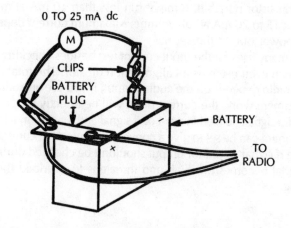

Fig. 3-4. Test leads with alligator clips can be very useful.

If the radio uses penlight batteries in holders, you can get your meter into the circuit with an adapter like the one illustrated in Fig. 3-5. Get a strip of heavy insulating paper (such as the fish paper used in electrical work) about one-half inch wide and two to three inches long. Cement a thin strip of brass (shimstock) to each side, and solder test leads to the ends of the strips. Be sure the brass strips are cut a little smaller than the insulator. To use this adapter, insert it between any two batteries in the string and connect the test leads to the milliammeter. It's easy to get the adapter in place if you lift the ends of two of the batteries, put the adapter between them, and then push them back into the holders.

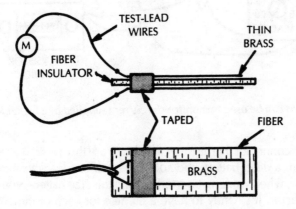

Fig. 3-5. This adapter can be employed to use a meter with a circuit powered by penlight batteries.

Check the service data to determine what the current drain should be. In a typical six-transistor portable, it might run less than 10 mA at minimum volume and about 15 to 20 mA at full volume. Maximum current depends on how much audio power output the set has.

There are many uses for the input current test besides checking bias, battery life, etc. You can even use it as an alignment indicator. The total current drain of a transistor radio depends on the audio output; if you use an audio-modulated signal for alignment work, the current drain will be directly proportional to the amount of audio signal. So feed in an i-f or rf signal and tune for maximum current. The volume control can be set to give a convenient amount of current. This setting and that of the rf signal generator output shouldn't be changed during alignment unless the signal becomes too high and threatens to overload the receiver or cause clipping.

TESTING POWER TRANSFORMERS FOR INTERNAL SHORTS

Many sets have "hot" power transformers. They might smell bad, have tar running out, etc., and in general show all of the symptoms of being hopelessly burned up. But are they? The big question is always this: "Is the transformer broken down internally, or is the overheating due to a short in the load circuits?" In all cases, we must know if the transformer itself is bad before we can make any estimate on the job. So, we check it first.

Remove the rectifier tube, or disconnect the silicon rectifiers, etc. In a few sets, you can do this by pulling the B+ fuse. In any case, be sure that the rectifiers are completely disconnected from the power-transformer secondary.

In a tube set you should open the filament circuit by disconnecting one wire from the power transformer. (You could pull all of the tubes, but that takes longer!) If the filament circuit is center-tapped, you'll have to open both supply wires. In all tests, make sure that there is no load on the power transformer.

Plug the primary of the power transformer into the wattmeter and turn it on. If the transformer is *not* internally shorted, you'll see a very small kick of the meter needle as the magnetic fields build up, and then the reading will fall back to almost zero. This reading, usually 2 to 3 watts at most, is the "iron loss" and is normal. If there is a shorted turn anywhere in the transformer, however, the wattmeter needle will come up to 25 or 30 watts. If the short is in one of the high-current windings, you'll see a full-scale, needle-slamming reading. This means that the transformer is *definitely* bad. Recheck to make sure that all loads have been disconnected. (Even two pilot lights can show a reading of about 5 watts.)

There is also a no-instruments-at-all test you can do on a power transformer. Hook up the transformer with no load on it, and leave it on for 5 to 10 minutes. If the transformer gets too hot, it's bad. A good transformer will get just barely warm running no-load. A badly shorted one will heat up and smoke.

SUBSTITUTE POWER TESTERS

Let's discuss the kind of equipment you can substitute for a wattmeter to get the same results. Since you need a "volts times amps" reading, you have to do your own arithmetic. The ac line voltage is usually specified as being 117 volts rms, but let's use 120 volts for simplicity. If you know the line voltage, measure the current and get watts by multiplying.

A 0- to 5-amp ac meter in series with the input, as in Fig. 3-6, gives this reading. For example, if the rating plate of the equipment under test reads 240 watts, our current should measure 2 amps. You can use a 0-to-5-amp ac ammeter of the panel-meter type for such measurements. Also, an adapter is available that makes a clamp-on ammeter out of a pocket VOM. It has a current transformer made so that the core can be opened up and clamped around either one of the ac wires. When the adapter is in place, the meter reads ac current on the ac volts scale.

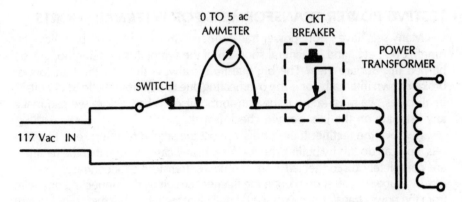

Fig. 3-6. A 0-to-5-amp ac meter in series with the input can be used to measure wattage.

This device is illustrated in Fig. 3-7. If you have one of these adapters, be sure to place the clamp-on core around only one of the wires; if both wires get inside it, you won't get any reading. On TV sets, you can usually get at one wire at the ac interlock, at a wire going to the switch, etc.

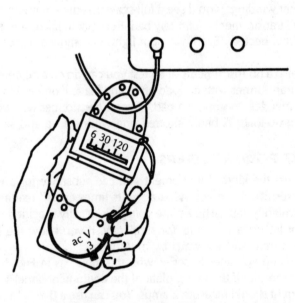

Fig. 3-7. This adapter permits ac current to be measured on the ac volts scale.

You can now read current, multiply it by line voltage, and end up with the wattage. But ac ammeters aren't common in service shops either; let's figure out another way to measure power. Use the ac voltmeter from the VOM. Get a fairly accurate 1-ohm resistor; the standard 5- or 10-watt wirewound types are an "automatic" Ohm's-law computer. Using E = IR, Mr. Ohm said that for every ampere of current that flows through a 1-ohm resistor, it drops 1 volt across it. So, you can read the ac voltmeter as an ac ammeter. In a 240-watt circuit, you'd read 2 volts (and therefore 2 amps) and get the same result as before. This technique is illustrated in Fig. 3-8.

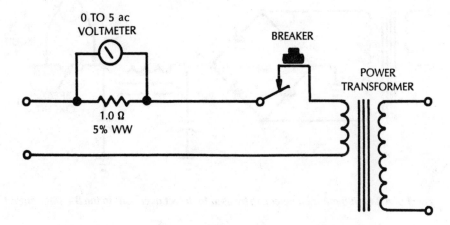

Fig. 3-8. Use Ohm's law to measure ac current on the ac volts scale.

The method just described is a valuable test for sets using small circuit breakers in the primary of the power transformer. If the breaker kicks out at odd intervals, check to see whether it is due to an intermittent overload in the secondary circuits or to an intermittent *breaker*. The kickout value of current is always specified on the breaker itself; for example, a typical unit might be rated "hold 2.2 amps; open at 2.5 amps." If we read the actual current in the circuit and find that the breaker is kicking out at 2.2 amps, we replace the *breaker* and that usually fixes the trouble!

FINDING OVERLOADS IN B+ CIRCUITS AND POWER SUPPLIES

After clearing the power transformer from suspicion in a fuse-blowing or overheating problem, we can put the loads back on one at a time and find out which one is faulty. One good test is to leave the filaments open and hook up only the B+. Check input power again. If this shows more than 30 to 35 watts, look out! If the current drain is as much as 75 to 100 watts with only the B+ circuits hooked up, there is definitely a short circuit in one of the branches.

Leaving the filaments off will raise the B+ voltage by taking off the normal loading. This test also can help to break down leaky parts and give us a nice, definite indication, such as a thin pillar of smoke. So do this test with care and with one hand on the switch!

If the transformer and B+ filter circuits check okay, you can read the B+ current drain by hooking a 0 to 500 dc milliammeter into the B+ filter output, as in Fig. 3-9. The normal current drain will always be given on the schematic; for example, 350 volts, 260 mA—and so on. If you get something like 300 volts at 290 mA, look out. Something is drawing more than its normal current.

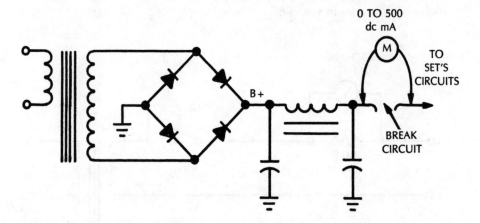

Fig. 3-9. A 0-to-500-mA millimeter can be used to detect overloads in the B+ filter output.

The excessive current can be traced by voltage readings in individual circuits, especially if they have their own individual dropping resistors. Look for the resistor with the greatest percentage of drop. For instance, if you have taps for 350, 250, and 150 volts coming off the B+ filter output, and the 350- and 250-volt lines read about 10 volts low while the 150-volt line reads about 50 volts maximum, the trouble is in the 150-volt line. For a quick check, disconnect the 150-volt line and see whether the others jump back up to normal (or probably a little above, because of the reduced loading). Having traced the trouble to a single circuit, you can chase down the short with an ohmmeter.

SETTING THE BIAS OF A POWER
TRANSISTOR WITH A CURRENT METER

Transistor bias voltages are critical. A change of only a small fraction of a volt can cause a transistor to cut off completely or draw a very high current—often enough to overheat the junction and destroy the transistor. If you suspect that a power transistor is drawing too much current, the only way to check it is to insert an ammeter and read the current directly.

Figure 3-10 shows how this is done on a typical single-transistor output stage. The bias is adjusted by means of a series resistor in the base return circuit. Open the bottom of the output transformer primary circuit, as shown, and insert a 0- to-1-amp dc meter. With this connection, read the collector current; its value is determined by the base bias. (You can consider this as either a voltage bias or a current bias; if you change current, you change voltage and vice versa, so don't be confused about it.)

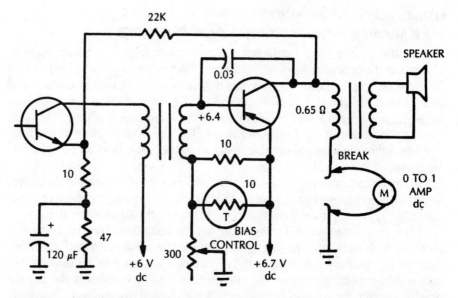

Fig. 3-10. A method for setting the bias of a power transistor with a current meter.

The circuit shown happens to be from a 6.3-volt radio, and the power output is not very high. The rated values for this circuit are as shown, and the bias control should be adjusted to give a collector current of 550 to 600 mA (0.55 to 0.6 amp).

The collector goes directly to ground through the primary of the output transformer. This winding has a very low dc resistance, so a dc voltage measurement here would be hard to read unless you had a meter that read millivolts. This explains why you take a *current* reading, which has more readable values. The actual current will be different on other sets, but the principle is the same.

For a given auto radio, details on how to adjust the output current are in the service data. Check these before you make the test, because auto radios differ in current values. The current in a given circuit depends on the type of transistor used and how the circuit was designed. You need the exact value so you can adjust for correct current in the set you are working on. The bias value sets the operating point of the transistor.

In two-transistor push-pull or transformerless output circuits, correct collector current is especially important. Most of these circuits work in class B: one transistor amplifies the negative half of each cycle, the other the positive. If the bias isn't exactly right, you get what is called crossover distortion at the point where one transistor stops conducting and the other starts. You can see this distortion with a scope; there will be a decided break or notch in the signal waveform where it crosses the zero line. Some sets use a combination bias; the output stage actually works in class A on small signals, shifting to class B on large ones. Bias is very important on these circuits. Combination-bias circuits are also found in hi-fi stereo amplifiers and are adjusted in exactly the same way as car-radio circuits.

USING CURRENT READINGS
TO BALANCE HIGH-POWERED AUDIO STAGES

Another case where current readings come in handy is when balancing plate currents in the output tubes of high-powered PA-system amplifiers. This should be done on amplifiers having 50 watts output or more. A typical circuit used in a 75-watt commercial amplifier is shown. Notice that there is a fixed bias supply, coming from a rectifier and filter in the main power supply. This bias supply will be about −75 volts dc; it is fed to the control grids of the final amplifiers through a pair of variable resistors so the two tubes can be balanced by changing the fixed bias.

Read the cathode current of each tube, which is the plate current at a lower-impedance (and lower-voltage) point. Although this is actually a current reading, take it by connecting a small resistor in series with each cathode circuit and then read the dc voltage drops across them with a voltmeter. (You can use this particular trick often as we go along; the right resistance makes a sort of automatic Ohm's-law calculator.) By making the resistance a known even value, you can take a voltage reading across it and interpret the reading in terms of the current flowing through the resistor. For instance, for a 10-ohm resistor, as in Fig. 3-11, every 100 mA through the resistor causes a 1-volt drop. Thus a reading of 1.5 volts indicates a current of 150 mA, and so on.

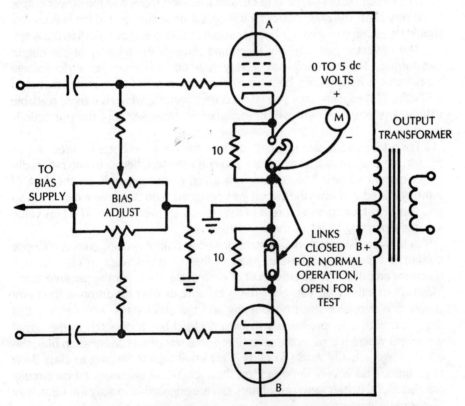

Fig. 3-11. Current readings can be used to balance high-power audio stages.

In this application, a shorting link is connected across each resistor when the circuit is in normal use. To make the test, both links are opened. The negative lead of the voltmeter is hooked to ground, and the positive lead is moved from one cathode to the other. The exact value of current is in the service data for each amplifier.

This test is usually made at low volume or with the gain all the way off. However, it can be used at a higher power level if you want to make a dynamic check for balance. The readings for the two tubes should be the same, or within a small tolerance. One can be brought down or the other one brought up. Make the grid bias more negative to cut down current or more positive to increase it. For high-level balance tests, a sine-wave signal from an audio generator must be used so that you have a fixed signal level at all times.

4
VOM and VTVM Tests

This chapter describes a wide range of applications for the technician's right arm: the multimeter.

MEASURING CONTACT RESISTANCE WITH A VOM

Once in a while you need to test a switch or relay contact to see if it is making good contact. A bad contact is annoying, especially if it is intermittent. However, there is one positive test: checking the resistance of the contact by reading the voltage across it in actual use. This test hook-up is simple, as illustrated in Fig. 4-1.

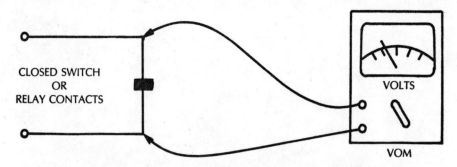

CLOSED SWITCH
OR
RELAY CONTACTS

VOLTS

VOM

Fig. 4-1. It is very simple to measure contact resistance with a VOM.

The contact resistance of a good switch should be zero. There should be absolutely no voltage drop across the contacts under full normal load. If the contact surfaces become dirty or burned, there will be resistance. Where there's resistance, there's heat, voltage drop, etc.

To check for contact resistance, set your voltmeter to the full line voltage of the circuit, and connect it directly across the contacts. The voltage can be either ac or dc; just set the meter to read the full voltage that the switch is breaking. With a sensitive meter, you'll see the full voltage across an open switch, even if the load circuits have fairly high resistance. Now, close the switch; the voltage should drop to zero. Turn the switch on and off several times to see if the voltage does drop to zero *every* time.

If there is any contact resistance, you'll see a small voltage reading when the contacts are closed. In most cases, this can be read with your initial meter setting. If you want to, turn the meter to a lower-voltage scale for a more accurate check, but be careful. The contact could be intermittent, and any resulting surge of voltage could damage the voltmeter (unless it is a VTVM).

With the meter set on the maximum voltage scale, tap the switch with the handle of a screwdriver to see if this makes any difference in the reading. It's a good idea to do this before going to a lower voltage scale with a VOM.

It takes only a very small voltage to show you that there is unwanted contact resistance. Even 0.5 volt means that there is some resistance between the contacts. This resistance will tend to increase in time, and the switch might eventually burn out. (Resistance makes heat; heat makes more resistance—it's a vicious circle.)

If the contacts are accessible as in some relays, you should clean them with very fine sandpaper or crocus cloth and a contact-burnishing tool or even a piece of cardboard. After cleaning, check again for zero resistance and zero voltage drop.

MEASURING VOLTAGE DROP IN LONG BATTERY WIRES

There might be some mystery when troubleshooting too much voltage drop in the supply wires. Take a typical case; a two-way radio transceiver is mounted in the trunk of a car and connected to the car's battery by long wires. If these wires aren't heavy enough or if there is contact resistance in a switch or relay, the radio supply will be low, and we'll have a "mysteriously" weak set.

The first step in such a case is to read the battery voltage at the radio power-supply plug. If the voltage is below normal, then go back to the battery itself and read its voltage. The radio must be turned on so you get a full-load reading; in most cases, it's desirable to key the transmitter to place maximum load on the power supply.

If the battery voltage is normal and there's a loss between the battery and the radio, the next step is to find out just what part of the supply circuit is guilty. Do this by taking more voltage readings—this time across the wires themselves. You'll probably need an extension lead for one of the voltmeter test leads because you have to take a reading between two points that are 18 to 20 feet apart! Don't worry about the length of the test lead. Because of the small current drawn by the voltmeter, voltage drop in the test leads will not affect the reading much.

Connect one test lead to the ungrounded post of the battery itself. If this is positive, as it is in most American cars, put the positive meter lead here. Make sure that you have a good, clean connection; a clip will be needed to hold the lead in place. Now put the other test lead on the positive (or "hot") wire of the radio power-supply plug, as at A in Fig. 4-2. Turn the radio on; there should be no voltage reading at all on the meter. Normally, you'll be able to see any significant voltage drop, since it must be from 1 to 3 volts to cause trouble.

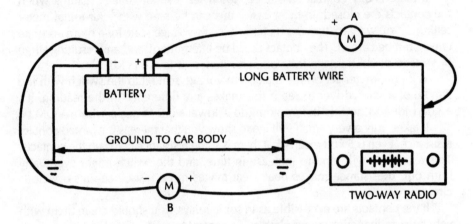

Fig. 4-2. Special techniques are used to measure the voltage drop in long battery wires.

If you see no voltage drop on the positive or hot lead, put the negative lead of the voltmeter on the grounded battery post. Then take another reading, this time to the ground point in the car's trunk or at the ground lead of the power supply plug (B). Again, you should see no voltage drop. However, if the ground path is not perfect—for example if there are painted joints between this point and the battery ground—you *will* see some voltage.

By doing these two tests, you can pin down the cause of any excessive voltage drop in the supply circuits. Don't forget that these wires also go through switches and relays, as well. To get away from possible voltage drops in automotive switches, most two-way radios and similar high-current equipment have relays that do the actual switching on or off; the car's ignition switch is used only to control this relay. Make sure the supply circuit does not go through the ignition switch itself; such switches are too light for this purpose.

IDENTIFYING UNCODED WIRES IN MULTICONDUCTOR CABLES

Identifying the wires in a large group of non-color-coded wires is difficult at best. You find this problem in intercommunication systems, where someone has bunched assorted wires without color-coding or identifying them. What do you do?

Get a 9-volt battery and a connection strip. Solder a pair of short test leads to the strip, with alligator clips at each end of the leads. You can hook a milliammeter in series with one lead, or use a pilot light (a 12-volt type). This is shown in Fig. 4-3. Now all you need is a 0-to-10-volt dc voltmeter and a set of wire tags. Numbered adhesive strips are sold by many electronics suppliers just for this purpose.

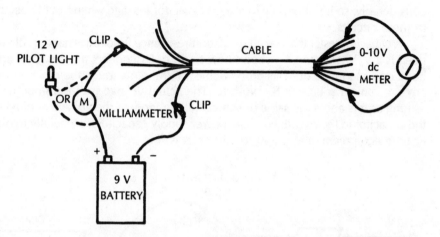

Fig. 4-3. This is a handy method for identifying uncoded wires in a multiconductor cable.

The testing takes two people. Find the two ends of the cable you want to identify. Put one person with the battery at one end, connect the battery to any two wires in the cable, and mark the negative wire No. 1 and the positive wire No. 2. At the other end of the cable, with your voltmeter, start checking between all possible pairs of wires for voltage. This is the hard part because you have so many possible combinations. If the wires have continuity, you'll eventually find the right pair. Mark the negative wire No. 1 and the positive wire No. 2. To signal the person at the other end, short the two wires together a couple of times. If he's using the pilot light, it will flash, indicating that you've found the first pair.

Your partner leaves the negative clip connected to the same wire (No. 1) and moves the positive clip to any one of the others. You leave your negative clip attached and search with the positive one until you find voltage on another wire. Mark this wire No. 3; then short it to No. 1 to flash the signal light. Your partner should have marked the positive wire No. 3 while you were searching and now moves the clip.

This technique also serves as a good short or leakage test. Even after you locate the wire with voltage on it, go on and check all the wires. See if you get a voltage reading on any of the others. You shouldn't, unless there is water in the conduit, an accidental short, etc. If there is a short, this method will reveal it quickly. Your partner's light will turn on the instant it's connected to the shorted pair.

TESTING CAPACITOR LEAKAGE WITH A VTVM

In checking small capacitors, find out if they are open, leaking, or shorted. Because of the way such capacitors are built, it's almost impossible for them to change value. A .001 μF capacitor is likely to remain a .001 μF no matter what happens to it. Rather than check the value as with electrolytics, do a condition test to determine whether the capacitor is functioning. A leakage of as little as 200 to 300 M in a high-impedance circuit can upset things considerably, so it's important to test for capacitor leakage when trouble occurs in these circuits.

A capacitor tester that has an insulation-resistance function is handy. All you need is a VTVM. Figure 4-4 shows how to do it. Disconnect the capacitor and hook one end to the dc-volts probe of the VTVM. Now touch the other end to any convenient source of B+ voltage. The actual voltage isn't too important, as long as it is about the same or somewhat higher than the voltage applied to the capacitor in the circuit. For large power supply capacitors, especially in older tube equipment, about 200 to 250 volts is fine.

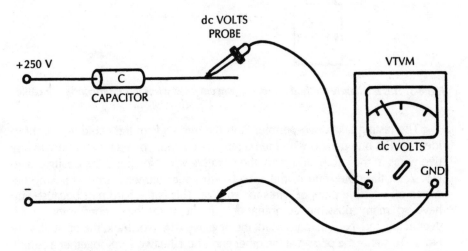

Fig. 4-4. A VTVM can be used to test capacitor leakage.

In modern solid-state (especially IC) circuits, much smaller capacitors are normally used, with rated working voltages of 25 volts or even less. Do not exceed the maximum rating of the capacitor under test or you could destroy it. For small, low-power capacitors, the ohmmeter battery in a VOM is sufficient. On the other hand, too low a test voltage might not give a readable result.

You'll see an initial kick of the voltmeter, caused by the charging current of the capacitor. If there is little leakage, this initial deflection will gradually ease off and the meter needle will go back to zero. The larger the capacitor value, the longer this takes. If you're in a hurry, you can touch the VTVM probe tip and ground with your fingers to discharge the capacitor faster.

Now, watch the meter needle. If it never reaches zero, or if it starts back up again after you have brought it to zero by touching the probe to ground as described, there is leakage in the capacitor. You're making a voltage divider that consists of the input resistance of the VTVM and the capacitor. Leakage current gives a voltage drop across the meter—and a voltage reading.

The actual voltage isn't too important; the fact that there is any voltage at all is what we're looking for. In coupling-capacitor circuits, even a 1-volt leakage can be too much. It causes a leakage of positive voltage from the plate of one stage onto the grid of the next stage, canceling out part of the negative bias voltage and perhaps throwing the tube onto the wrong part of the curve and causing a severe distortion.

With transistors in similar circuits, capacitor leakage is even more important. It takes only a tiny fraction of a volt to make a transistor cut off or run wild. The latter can cause overheating and avalanching and can destroy a transistor quickly.

For best results, check any new capacitor before putting it into circuit. Such a check can save a lot of time if the new capacitor does happen to have a little leakage! We have a tendency to automatically assume that a new part is good. This isn't true in *all* cases; check first, and you'll be sure.

MEASURING VERY HIGH RESISTANCES

You may have occasion to check high-value resistors that are far beyond the range of your ohmmeter scales. There is a way to do this—in fact, there are several ways.

The easiest is with a resistor of the same value or one as close as possible. If you don't have an exact duplicate, you can connect several resistors together temporarily to get the necessary value. For example, for a bleeder resistor used in color TV focus-rectifier circuits—66 M, hook three 20 M resistors and a 6 M resistor in series; or use three 22 M resistors; etc.

For the test, connect the combination in series with the suspected resistor and connect the whole string across a source of voltage. The B + voltage of the TV set is handy, but you can use any voltage, even the ac line if you want. Take a voltage reading across one of the resistors, as shown in Fig. 4-5, and note the reading. Next, check the voltage across the other resistor.

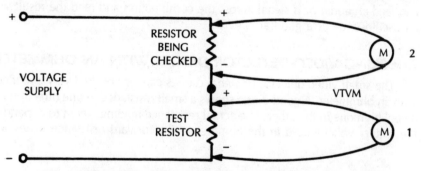

Fig. 4-5. Very high resistances can be measured with this method.

Because you have deliberately made a divide-by-two voltage divider, the two readings should be almost the same if the resistor under test is okay. If the difference is not more than 10 percent, the resistor is probably good. (The amount of difference you should tolerate depends on the resistors used; if they are 5 percent resistors, then cut down your tolerance to this amount, and so on.)

If you don't have the right resistors values, you can still use some that will make the total resistance of the voltage divider come out at a value that is easy to interpret. For example, with the 66 M resistor, you could use a single 22 M test resistor. Your whole circuit would then offer 88 M (66 + 22); your voltage reading across the 22 M resistor would be one-fourth of the total voltage. If you used 200 volts, you'd read 50 volts across the 22 M and 150 volts across the 66 M.

Note that this method—checking the voltage across each resistor in turn—takes the meter resistance out of the picture only if all resistors are equal. If you had to figure the actual voltage present across each resistor, you'd have to figure out the shunting effect of the meter resistance.

CHECKING DIODES BY THE BALANCE METHOD

You can use the method described for checking very high resistances to check diode back-resistance, if you want. Ordinary diodes, such as afc, video and sound detector, etc., can be checked with an ohmmeter, as will be shown in a later unit. However, if you run into special units, such as the high-voltage diode rectifiers used in focus and boosted-boost circuits in color TV sets, the ohmmeter simply won't reach the resistance range needed.

The duplicate-and-balance method shown in Fig. 4-6A is the easiest. Connect the suspected unit in series with another one just like it, and feed a voltage to the combination for equal voltage drops across the two units.

You can also use this method for testing low-voltage diodes. This technique is useful for checking balance on afc diodes, ratio-detector diodes, and diodes used in any of the circuits of FM multiplex receivers and decoders. With these low-voltage units, connect the diodes in series and add a series resistor to keep the current within safe limits (see part B of the figure). Now hook a small test voltage across the combination and check the diode voltage drops for balance.

As a general rule, if any diode—from the 5,000-volt focus rectifiers down to one of the tiny video-detector diodes—is really bad, the defect will show up in the front-to-back resistance ratio, which can be measured by this test. You can feed an audio or rf signal across the combination and read the result with an oscilloscope or a good ac VTVM.

CHECKING VIDEO-DETECTOR DIODES WITH AN OHMMETER

The small video-detector or signal diodes can be checked very accurately with an ohmmeter. A good diode shows a small resistance in one direction and a very high one in the other. The actual resistance readings seem to depend on the battery voltage used in the ohmmeter. The forward resistance varies with voltage and diode types but will probably be in the neighborhood of 100-200 ohms. The reverse-bias measurement should read several hundred thousand ohms.

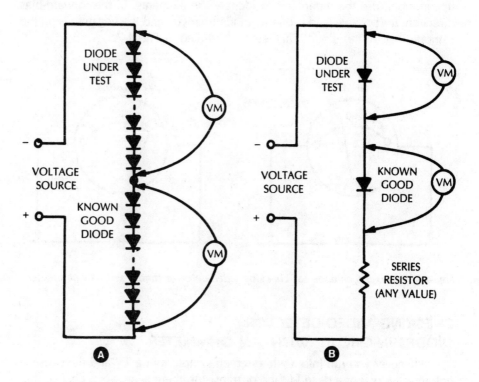

Fig. 4-6. Diodes can be checked with the balance method.

Practically all VTVMs use a low-voltage battery in the ohmmeter; the average voltage is about 3 volts. Most VOMs use a 1.5-volt battery on the low-resistance ranges, but use up to 22.5 volts on the highest resistance scale in the megohm range. Diodes should be tested on low range—a scale of 0 to 2,000 or 0 to 5,000 ohms—because such a range will use a low-voltage battery.

Find out which of your ohmmeter leads is positive, that is, which one is connected to the positive side of the ohmmeter battery. Check yours with another voltmeter and mark the positive lead. A voltmeter check also tells you exactly what voltages are used in your ohmmeter on various ranges.

If the positive ohmmeter lead is placed on the cathode of the diode and the negative on the anode, you get a high-resistance reading. Reverse the leads for a low-resistance reading. The ratio between the two should be very high for best performance in video detectors and similar circuits.

Actually you don't have to bother with polarity. With a high resistance one way and a low resistance with reversed leads, the diode is okay. But with an unidentified diode or one so small that you can't tell what the markings are (not uncommon), use the ohmmeter to identify the elements. In the forward-bias direction, the positive lead is on the anode (triangle) and the negative is on the cathode (bar) of the diode. This test is illustrated in Fig. 4-7.

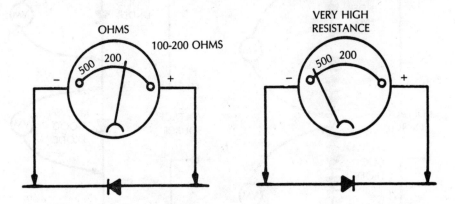

Fig. 4-7. This is a technique for checking video-detector diodes with an ohmmeter.

CHECKING VIDEO-DETECTOR DIODES IN-CIRCUIT WITH AN OHMMETER

Whenever you run into a white-screen symptom in a TV and the cause is not an agc block or a dead i-f tube or transistor, there is always a chance that it could be a bad video-detector diode. These diodes are usually contained in a small can, and they shouldn't be unsoldered and resoldered too many times. We can make a definite and reliable test without moving a wire: the ohmmeter will tell us from outside whether the diode is good or bad.

Figure 4-8 shows a typical video-detector circuit. The i-f signal comes to the diode from a secondary winding on the last i-f transformer; the other end of the winding is grounded. (This is a series-diode detector, but shunt detectors can be checked in the same way.) The coils are peaking chokes, which give the circuit a wideband response, and the video-diode load resistor is the 3,300-ohm unit connected from the diode output circuit to ground.

To check the diode, first hook the negative lead of our ohmmeter to the video coupling capacitor as shown. That way you can read the dc path through the peaking coils and diode to ground. The positive ohmmeter lead is connected to ground (*1* in the figure). This is the reverse-bias direction, so you should read nothing but the load resistor plus the small resistance of the two peaking coils—actually, 3,318 ohms (but if you get 3,300 ohms, it's okay).

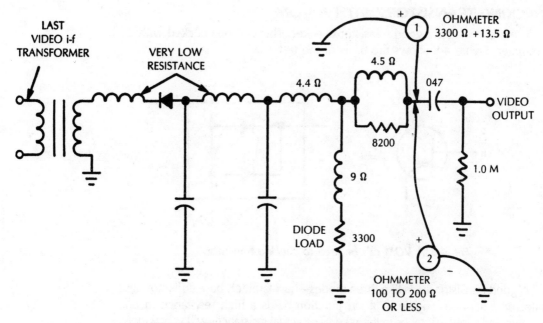

Fig. 4-8. This is a typical video-detector circuit.

That's one check. Now, reverse the leads—positive to the coupling capacitor, negative to ground (2 in the figure)—and read the diode in the forward-bias direction. Our path is through the low forward resistance of the diode with the 3,300-ohm resistor in parallel, so the measurement should read something like 150 ohms. Simply ensure there is a good, high ratio between the two readings.

If the reading is infinity on both tests (no reading at all), the diode or one of the peaking coils is open. Doesn't matter which; you're going to have to take the circuit apart to find out which one and fix it. Check each of the coils for continuity before desoldering the diode; such coils sometimes open because of corrosion on the fine wires. If you have a coil shunted with a resistor—such as the last coil in the circuit shown, which is shunted by 8.2K—the value of the reading can tell you which coil to suspect. For example, if you read 8.2K + 3.3K from grid to ground (in the high-resistance direction) you'd know the last coil was open, and check it first.

If the reading is the same in both directions, and very low, the diode is shorted. You can figure what the correct resistance of the circuit should be by totaling the resistances of the peaking coils and diode load resistor, etc. If you have any doubts, you can lift the ground end of the diode load resistor, thus eliminating the shunt path, and read only the diode's resistance.

In shunt detector circuits, where the diode is connected from the transformer output to ground, you can also trace the dc path for the ohmmeter reading and catch an open or shorted diode in the same way.

CHECKING TRANSISTORS WITH A VOM

The VOM makes a good transistor checker. This can be checked with an ohmmeter. Figure 4-9 shows the basis of the test.

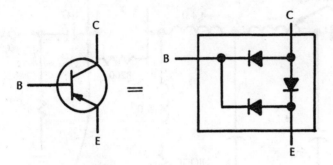

Fig. 4-9. A VOM can be used to check a transistor.

A good transistor reads like three diodes—base-emitter, base-collector, and collector-emitter—as shown. A good junction reads a high resistance in one direction and a low resistance in the other. The actual resistances will vary widely, depending on the battery voltage used in your ohmmeter, the type of transistor (silicon, germanium, etc.), and its rating. Look for the ratio between the two readings for each diode; this ratio should be high. Typical values might be 200 ohms in the forward direction and 50,000 to 75,000 ohms in the reverse.

Use the low-resistance scale on your ohmmeter, particularly in a VOM. This scale usually has a low-voltage battery (1.5 to 3 volts) and won't damage the transistor. The high-resistance (megohms) scale can use up to 22.5 volts, which could damage low-voltage transistors.

To check each diode, take one reading, and reverse the leads to read the resistance in the other direction. If the reading is the same in both directions (100 ohms or less), that junction is shorted. For a valid test, the transistor must be disconnected; there are often very low resistances, such as coils, low-value resistors, etc., connected across the junctions. If you get a reading of infinite resistance in both directions, the junction is open; this defect is fairly rare. If you read something like 100 ohms one way and only about 500 ohms the other way, the transistor is probably leaking across that junction and should be replaced; a 5-to-1 ratio is too low.

VOLTAGE AND RESISTANCE TESTS

Voltage measurements in radio and TV work are helpful. They can also be misleading—that is, unless you know what you are reading, where you're reading it, what you're reading it with, and what the reading should be! Again, all tests are comparisons against a standard, and in the case of voltage measurements the standard is the operating voltages given on the schematic diagram.

In low-impedance circuits such as the power supply, you can use any kind of voltmeter. The circuit can supply so much current that the type of meter used makes no difference. However, when you get farther along in a piece of equipment into some of the high-impedance stages, the meter impedance plays a more significant part.

Figure 4-10 shows a good example. This is one of the popular tube circuits used in many older TV sets as a sync separator and agc stage. Note the values of some of these resistors. In some cases, you need to read a dc voltage through a resistance of 10 or 12 M. (All of the coupling capacitors, etc., that feed signals into this circuit have deliberately been left out because you are only concerned with the dc voltage relationships in this type of circuit. However, the ac signal does make a great deal of difference, not only in the output, but also in the dc voltages themselves.)

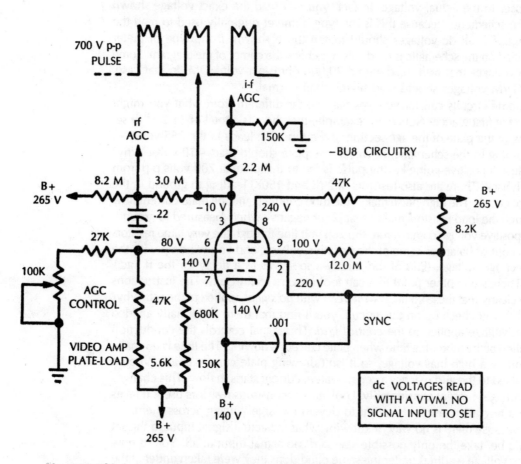

Fig. 4-10. This is one of the popular circuits employed in a number of older TV sets as a sync separator and agc stage.

Yes, this *is* a complex circuit. But you can check it rapidly if you know what to look for and the meaning of what is seen. The diagram shows a set of voltages on the tube elements. Suppose you take a set of readings with a VOM. They will all be different from the values called for on the schematic. Why? Because we didn't read the instructions, that's why. Look in the corner of the schematic; it says very plainly, "dc voltages read with a *VTVM*."

For instance, if you put a VOM on a 0 to 100-volt scale (20,000 ohms per volt) and check pin 9, you won't read 100 volts as the schematic calls for. Our meter resistance on this scale is $20,000 \times 100 = 2$ M. Look at the size of the dropping resistor—12 M! If you shunt only 2 M from this point to ground, you load down the circuit so that the voltage drop across the 12 M resistor is far above normal. The result is a "wrong" voltage.

If you use a VTVM, with an 11 M input resistance, you get a reading that is closer to the actual voltage. In fact, you can read the exact voltage shown on the schematic, because this is the type of meter originally used to read the voltage. So, all dc voltages should match those shown, *if* the type of meter specified on the schematic is used. Always check the corner of the diagram. Some have voltages that *were* read with a 20,000 ohms-per-volt VOM. In that case, all VTVM voltages should read higher than normal.

Some circuits call for voltages that are far different from what you might expect at first glance. Notice, for example, that there is a total of 11.2 M in series with the plate of the agc section of the tube that leads to the 265-volt B+. According to the schematic, however, the plate should read −10 volts. Why? Because a positive-going keying pulse is fed to this plate at 700 volts p-p from the flyback. There are also resistors (2.2 M and 150K) leading to ground in the i-f agc circuit. The high pulse makes the tube conduct, and the plate-current flow through the load resistors makes that point negative when measured to ground. The positive voltage coming from the 265-volt line through the very large resistors bucks out or balances some of the negative voltage, and you wind up with the correct agc voltage. (Part of this circuit also serves as a delay for the rf agc.)

There's one other point that can fool unwary technicians. The instructions state clearly that these voltages were read with *no signal* applied to the set. Why? Well, if you check up on this circuit, you'll find that there is normally a video signal voltage applied to the *control grid*. This signal controls the conduction duration of the tube—the time when plate current is flowing. The tube is normally cut off by a high bias voltage, so it isn't drawing plate current at all. When a signal fed to the grid reaches a certain value, current starts to flow. This changes the voltages present considerably. Look at the monstrous resistors used; it takes only a few microamps of current to develop a large voltage across them.

Because there is no way of knowing what the actual signal input to the set would be, take the only possible standard: no signal input at all. You're now taking voltage readings under the same conditions they were taken under at the factory.

Voltage and Resistance Circuit Analysis

In actual field servicing, operating voltages provide our best clue to trouble by *changing* from the normal (becoming either too low or too high).

Whenever you find a tube or transistor with an incorrect voltage on any of its elements, check the complete circuit to find out why. After the voltage analysis, turn the set off and make a resistance check of all parts in the path in question, all the way back to the supply point.

For a typical example, refer back to Fig. 4-10. Suppose that either the 8.2 M or the 3 M resistor has increased in value. This would reduce the positive voltage on the plate of the tube, making the plate go too far negative. This in turn would make the agc too negative, and the controlled stages would be cut off—an agc whiteout. If the 0.22 μF bypass capacitor leaked, this would add a low-resistance shunt path at the junction of the two large resistors. Once again, there would be too little positive voltage to balance the circuit, and the same symptoms and a whiteout would occur.

How do you pin down the faulty part in such circuits? Measure the values of *all* resistors in the path where the fault may be. If there is too much negative voltage at the agc plate and too much positive voltage at the junction of the 8.2 M and 3 M resistors, but the 265-volt supply checks normal, then suspect the 3 M resistor of being open or greatly increased in value. If there was too little (or no) voltage at this junction, suspect the 8.2 M resistor of being open.

To get a correct reading of these resistors, open the circuit. Lift one end of a resistor before making any resistance measurements across it. To read the exact value of the 2.2 M filter resistor from the plate to the i-f agc, lift one end of the 150K shunt resistor to ground. If not, there would be at least two paths: one through the resistor and the 150K resistor to ground; the other, back through the big resistors to B+, which usually measures about 20,000 ohms to ground because of the leakage resistance of the electrolytic filter capacitors.

Transistor Circuits

In transistor circuits, you have the same basic problem with shunt paths, but with very low resistances. Figure 4-11 shows a typical class-A amplifier stage with an npn transistor. You might find such a stage in the preamplifier circuits of a hi-fi audio amplifier or in some similar type of equipment.

The signal here is developed across the load resistor R_L. This resistor is of such a size that half the supply voltage will be dropped across it, and the plate current can swing above and below the operating point an equal amount; this is what makes it a class-A stage. With an 18-volt supply, there's a 9-volt collector potential measured to ground. (If this were a pnp transistor, it would be the same thing, but the collector voltage would be negative.)

Notice the voltage divider network, resistors R1 and R2. They are proportioned so the base will have the right bias voltage for the type of transistor used—here, 0.69 volt. But the schematic says very plainly: 1.69 volts! The emitter is 1 volt above ground, so the actual base-emitter voltage is $1.69 - 1.0 = 0.69$ volt.

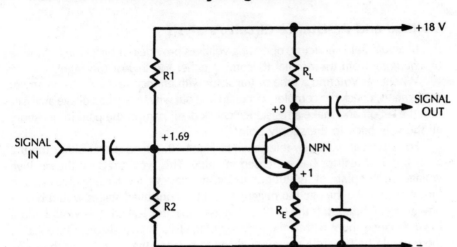

Fig. 4-11. A typical transistorized class-A amplifier stage.

In the tube circuit of Fig. 2-1, recall that both grid and cathode of the tube read 140 volts from ground. What is the actual grid-cathode voltage? That's right: 140 − 140 = 0! In case you were wondering about the statement we made that the tube is normally cut off, look at pin 6. This pin is a second control grid in this type of tube and actually has a −60-volt bias: 80 − 140 = −60.

In circuits like this, there are shunt paths again. You can't read the resistance of either R1 or R2 correctly unless you open the circuit. The resistance of the base-emitter or base-collector junction is in parallel with these resistors. Take out the transistor or lift one end of the resistor to get a correct resistance reading. If the voltages are off-value in such circuits, don't try to pin the trouble down by doing in-circuit tests with an ohmmeter. Break the circuit up into small sections and check each part by itself. Wrong bias on the base of the transistor could be caused by a shorted transistor or by incorrect values on R1, R2, or even R$_E$.

The voltage-resistance test sequence is probably the most useful in any kind of electronics work. Properly applied, it can give you a great deal of information in the least possible time. Using improper methods can lead you farther and farther away from the real trouble. The nonexistent defects aren't faults at all, but simply incorrect indications caused by the wrong test equipment!

In general, here's the test sequence: Find the defective stage by signal tests, then check its operating voltages. If any of these is off value, turn the set off and find the cause of the incorrect voltage by tracing the supply circuit back to its source. Measure all resistors in this path, including the resistance of coils, chokes and anything else that is a part of the dc path back to the supply. Learn to trace out a circuit and follow it back to the source of power, wherever it is. Start at this point, making sure the *source* voltage is correct for the very first test. You would be surprised to know how many technicians overlook this obvious and extremely simple test! By beginning at the supply and following the circuit until you come to an abnormal indication, you can pin down the trouble in the least possible time.

VOLTAGE REFERENCE POINTS: THE GROUNDED PLATE

What would you think if you were reading dc voltages in a TV set and ran into something like the circuit shown in Fig. 4-12? The circuit is perfectly normal as far as operating voltages are concerned. The plate of the tube has the rated 150 volts on it if you measure the plate voltage from the proper reference point—the tube's own cathode. Look at the smaller circuit in the figure; there is an amplifier tube, its load resistance, and the power supply. Note that this circuit has no connection to ground. If you measure the plate voltage from the cathode, you should get a high positive voltage. As far as the performance of the tube in amplifying a signal is concerned, attach a ground at any point and it'll make no difference.

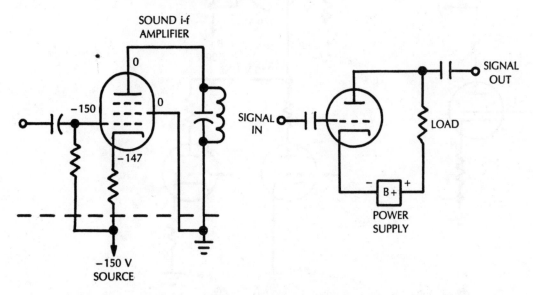

Fig. 4-12. This circuit can be confusing because it has a grounded plate.

This circuit is most often found in old sets, but that's not saying you will never see it used in newer ones. It is common in transistor circuitry, especially in the totem-pole or multiple-stacked direct-coupled circuits. Here, you must remember to use the proper reference point to get the exact voltage on any transistor (or tube). In the popular common-emitter circuit, the emitter is the reference point; base and collector voltages are measured with respect to the emitter. The bias on a transistor, of course, is the base-to-emitter voltage.

STACKED STAGES

Circuits with stacked stages have confused many technicians because of the unusual dc voltage relations. Let's consider the tube version first, because the involved theory is a little clearer and more direct with tubes.

Dealing with a circuit like the one shown in Fig. 4-13 is easy enough when you use the right reference point for each tube—the tube's own cathode, remember? The basic stack circuit is shown in Fig. 4-13A. Two identical tubes are hooked up in series. With 300 volts on the plate of the top one and the same bias on both, they'll divide the voltage. Voltage readings to ground are shown.

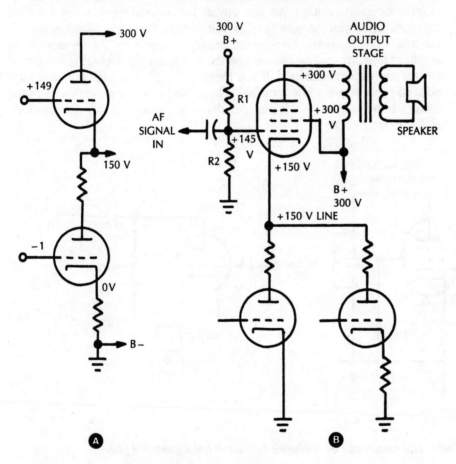

Fig. 4-13. Dealing with a circuit like this one is easy when the right reference point is used.

The grids have a bias of −1 volt each. Reading from ground, there's 150 volts on the cathode of the upper tube, and 149 volts on the grid; 149 − 150 = −1. In actual circuits, a resistor network is used to set up this bias, but we have left it out so that you can see the basic circuit more easily. A bias of −1 volt is also on the lower grid, with the cathode grounded.

Figure 4-13B shows an actual circuit used in many TV sets. The audio output tube is the top one, and several other tubes are fed from the voltage developed at its cathode. Its grid bias is fixed—notice voltage divider R1 and R2 with the grid connected to the junction.

The audio tube is a power pentode, with a plate current of 40 to 50 mA. You can feed any other circuits you want to, as long as the total of their plate currents doesn't exceed this amount. Since these circuits are usually voltage amplifiers, sync separators, etc., they don't draw a lot of plate current, and many stages can be fed from this source. In common practice, the voltage supplied to these stages will be from 130 to 150 volts, and on many schematics the circuit carrying this voltage is called the 150-volt line.

Notice the similarity to the basic circuit of Fig. 4-13A. We have deliberately made the voltages the same. (Actually, we have left out the 5-volt drop across the plate-load resistor. The plate of the audio tube usually reads about 295 volts.) You'll find other voltages used, but the principle is the same in all circuits of this type.

The audio output stage, with fixed bias on the grid of the pentode, has a certain amount of voltage-regulator action. The fixed bias holds the grid voltage almost constant, so there is little fluctuation in the average plate current. The audio fluctuation is filtered out in the cathode circuit by large capacitors that aren't shown here.

The actual bias on the tube is −5 volts; 145 − 150 = −5. This bias is determined by the ratio of the voltage-divider resistors and by the B+ supply voltage fed to the divider.

Many other circuits can be fed from the 150-volt line. For example, one TV model feeds the sync separator, video i-f, agc tube, and the CRT brightness control from the 150-volt line. Troubles in any of these stages could be due to something in the audio output stage. Therefore, it's a good habit to look at the schematic to see if the set you are working on uses stacked stages. If you don't have a schematic, measure the cathode voltage of the audio output tube. If this is 150 volts or so, you can be fairly sure that there is a stacked circuit involved. If the cathode is at ground, it's not a stack.

You'll find stacks in other places. The 6BQ7 in tuners is basically a stacked-stage arrangement. Also, many sets use stacked video i-f's; the first two tubes will be stacked, with 300 volts on the input plate and 150 volts on the second.

TRANSISTORS IN STACKED CIRCUITS

In transistor amplifiers, the equivalent of stacked-tube stages in various output-transformerless and direct-coupled stages is shown in Fig. 4-14. As in the case of tube circuits, the term stacking refers to the dc voltage distribution between the two (or more) transistors. The circuit shown in Fig. 4-14 is used in a commercial stereo amplifier and is a section of only one channel. It is a direct-coupled driver and OTL (output-transformerless) output stage. The circuit works in class B, with one transistor handling the positive half-cycles and the other handling the negative; the speaker is connected to their midpoint and ground.

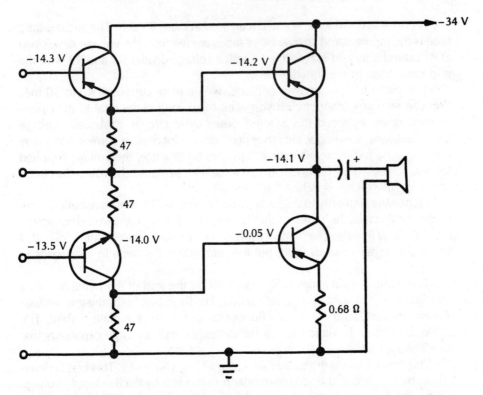

Fig. 4-14. This stacked transistor circuit is from a commercial stereo amplifier.

Notice that the maximum power supply voltage is connected to the collectors of the upper transistors and that the emitter of the upper output transistor does *not* go to ground, but to the collector of the lower transistor. The emitter of the lower transistor is connected to ground, which is the positive point of the power supply. (Remember, in any transistor circuit, the polarity of the power supply is determined by the type of transistor used—npn or pnp.) You can have a supply voltage of − 34 volts for one type or + 34 volts for the other and it'll work out the same. The main interest now is in the proportion of the supply voltage taken by each transistor, because this tells if the circuit is working properly.

Notice that the voltage does not split in the middle, as it usually does when similar vacuum-tube stages are stacked; maximum supply voltage is − 34 volts, but the midpoint of the circuit has only − 14.1 volts. The midpoint value must be checked on the schematic so you know what is correct.

Incidentally, transistor circuits do not have the rather wide voltage tolerance that most tube circuits have. If the schematic specifies 14.1 volts, it doesn't mean 14.3 or 13.8—it means 14.1 volts. In most transistor circuits, a change of only 0.2 volt can cut a transistor completely off or drive it to full conduction and overheat the junction. Learn to measure these very small voltages with extreme care. Measuring the collector current is a very good way to check the bias voltage. There is a much greater meter deflection, and it is usually easier to read on shop-type equipment.

Troubles in circuits like these are the same as in other stacked circuits; resistors that have changed in value, transistors with leakage between elements, inter-element shorts, etc. The emitter-base voltage (the bias) is a good place to start measuring if you suspect trouble. Use a sensitive and accurate VOM. A digital meter would be helpful here. Read the voltage directly from emitter to base. If this voltage is in tolerance, the transistor is probably okay. If it is off by more than 0.2 volt, then there is very likely something wrong in that stage.

Note the direct coupling used in the driver transistors (the left-hand pair in Fig. 4-14). This is a very common circuit and has the same type of voltage distribution as the push-pull circuit. The available supply voltage will divide between the two in a ratio determined by the type of transistor and the bias on each one. Consult the schematic diagram to verify the correct voltage.

5

Oscilloscope Tests

A typical service-type oscilloscope will not read dc voltage, dc current, ac current, or even ac voltage—directly. (Please note I said *directly*. You can read ac voltage on a scope, but only by *comparing* the vertical deflection to the vertical deflection of a known ac voltage from a calibrator.) Despite this, the scope is the handiest instrument in the whole shop. It can do something no other instrument can—read *signals*. In one of the most simple tests—using a test probe to touch one point and then another in a circuit—the scope can immediately and definitely depict if a given stage has any gain. Because a lot of time is spent trying to locate certain stages with insufficient gain, the scope can be a real timesaver. In many cases, it is the *only* instrument that can give the necessary information.

An ac voltmeter can determine the ac voltage in a circuit, but the scope is the only instrument that determines whether the voltage is hum, vertical sync, video, audio, or some other type of waveform. By comparison with a voltage standard (calibrator), it also depicts amplitude.

Make gain checks with ease: feed in a test signal of the frequency normally used in the circuit and check the signal levels at the input and output of any amplifier stage. If you read a vertical deflection of 1 unit on the input and 8 units on the output, the stage has a voltage gain of 8.

The scope is the closest thing to a really universal test instrument. It works with tubes, transistors, printed circuits, integrated circuits, or any other kind of circuitry with equal ease. Why? Because it reads signals, the one thing all circuits have in common. No matter how a circuit is built or what voltages it uses, it handles signals—amplifies them, clips them, shapes them, or does something to them—and you can check these actions with the scope. Use the scope in radio and TV work as often as the VOM or VTVM; it really speeds up the work.

Let's examine a few typical scope patterns—learn to recognize and interpret them. Like most tests, the important question isn't "What have we here?" but "What does it *mean?*" As an example, consider the scope display shown in Fig. 5-1.

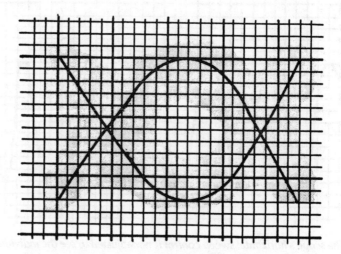

Fig. 5-1. A scope signal with the sweep set off frequency; however, this information can be useful.

Notice that the pattern height covers 12 small squares on the calibrated screen. The vertical deflection represents an ac signal, but the scope sweep is set off-frequency, so you can't see the individual cycles of the signal. It could be a sine wave, square wave, or anything.

However, this scope image can be used. Suppose you're feeding an audio signal into the input of an amplifier and touch the probe to the circuit. If you see this pattern on the input of a stage (tube grid or transistor base) and move the probe to the output (tube plate or transistor collector) and see the same pattern (height and all), it means that the stage has a gain of 1.

What this signifies depends on what you're testing. The scope reads *voltage.* If testing the preamplifier stage of a tube amplifier, this stage should have a very high voltage gain (up to 50). However, if testing a transistor preamp stage, then a 1:1 voltage ratio might be perfectly normal; such transistor stages often have a 1:1 voltage ratio but a large *current* gain that gives the required gain in signal *power.* For a voltage ratio of 1:0.25, it's trouble.

If looking for distortion, there's no need to see the individual cycles. Feed in the signal and touch the probe to any point in the circuit where the signal can be seen (preferably at the input), so that you know what the test signal looks like before starting. Then adjust the scope's horizontal sweep to see the individual cycles, as illustrated in Fig. 5-2. Adjust the sync-lock control of the scope to stop the pattern. This is a fairly good sine-wave pattern, but spread it out a little to get a better look at the waveshape. To do that, use a faster sweep or turn up the horizontal-gain control.

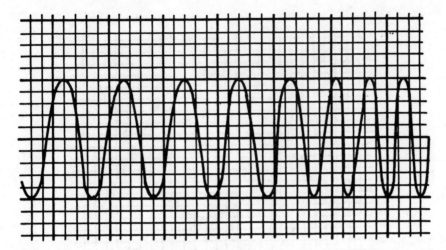

Fig. 5-2. The scope's horizontal sweep control is nonadjusted to see the individual cycles.

If the waveform shows distortion and you suspect that there is signal clipping somewhere in the amplifier, trace the signal through the amplifier from the input until a pattern similar to the one shown in Fig. 5-3 appears. Figure 5-3 shows a decided clipping on one-half of the sine-wave signal. By checking the circuit, you can tell which part is most likely to cause this distortion. In certain circuits, this could be the correct waveform. In a transistor class-B output stage, each transistor carries one-half of the sine-wave signal. However, if the output at the speaker looks like Fig. 5-3, only one-half of the circuit is working so look for trouble there.

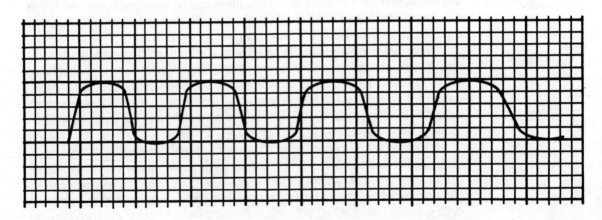

Fig. 5-3. This sine-wave signal is severely clipped on one half-cycle.

You can identify positive and negative halves of a sine wave. You only need to know which way the scope deflects on a positive-going signal. Most scopes have a reversing switch that will invert the pattern, and this switch must be in the NORMAL position before the test. Hook a 1.5-volt battery to the vertical input and see which way the beam jumps. Most service scopes are set up so a positive voltage makes the beam go up with the reversing switch in the normal position. In an ac-coupled scope, the beam will jump up or down, and then come back to the original position. In dc-coupled scopes, the beam will be permanently deflected by a specific amount.

By setting the switch so the positive voltage makes the beam deflect up, you can identify the positive and negative halves of the signal waveform. This is useful in the class-B transistor-output stages just mentioned. You can also tell where the zero line is. The zero line is the position where the beam rests when there is no signal input. By adjusting this to the center line on the calibrated screen, you can see how much of a waveform is positive and how much is negative.

It is not imperative for a scope pattern to be clear and sharply focused to be useful. In the horizontal-oscillator tests given later, we can use a pattern like the one shown in Fig. 5-4. What we're looking for in this case are horizontal sync pulses in a TV signal. The figure shows three of them; the third is at the extreme right of the pattern. (Our scope sweep, therefore, is running at 15,750 Hz divided by 3, or 5,250 Hz. Use this setting to make comparison tests anywhere in the horizontal sweep circuits.) Don't waste time trying to make picture-book patterns when all you need to know is the number of cycles of signal that can be seen on the scope.

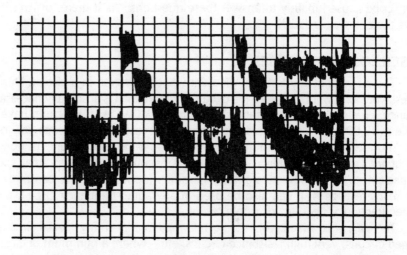

Fig. 5-4. This scope display includes the horizontal sync pulses in a TV signal.

CALIBRATION FOR DIRECT MEASUREMENTS

The height of the pattern on the scope screen depends on the setting of the vertical gain control. A 2-inch pattern means nothing, *unless* you have previously set the vertical gain for a 2-inch deflection when a known voltage (1 volt p-p, or 10 volts p-p) is applied to the input. One way to set the gain is with an external voltage calibrator, which is just a tapped transformer with an accurate ac voltmeter connected across the output terminals. The scale on these meters is usually calibrated in rms, peak, and peak-to-peak voltage. A given deflection can be read as any one of three, depending on what we want it to mean in a particular test. *Unless a special probe is used, a scope always displays peak-to-peak voltage.*

Many scopes have built-in voltage calibrators. Some have variable calibrating voltages with a volt-reading knob. Others have a regulated 1-volt p-p output. By using the calibrated step attenuator, set the scope for 1 volt on the lowest (most sensitive) range, and turn the attenuator to ×10 and read 10 volts for the same deflection; to ×100 and read 100 volts for the same deflection; etc. Check the instruction book for your scope to see how this attenuator is marked. Some are marked in gain and others in attenuation. They do the same thing, the markings are just different.

If you don't have a scope calibrator, you can use any kind of variable ac voltage, measuring the value with an ac voltmeter (remember that the standard ac meter reads *rms* voltage). The filament voltage of your tube tester is a very useful calibrator, since it runs from 1.1 volts to 117 volts. This, too, is in rms values, and you'll have to use the conversion formula: 10 volts rms = 14.14 volts peak, and 28.28 volts peak to peak. If you want a very rough peak-to-peak measurement, just figure that the peak-to-peak is three times the rms voltage.

As you go through the various scope tests to speed up electronics servicing, the scope is used mainly to answer these questions: "Is it there, or isn't it; if it is there, how big is it?"

OSCILLOSCOPE PROBES

The purpose of any test is to discover what's going on in a circuit. During a test, you don't want to disturb the circuit any more than is absolutely necessary. Hence, I prefer the high-resistance VTVM to the VOM. A VOM could have as little as 40,000-ohms input resistance (on a 2-volt scale, 20,000 ohms-per-volt). The most sensitive instrument in the average shop is the oscilloscope; its normal input has a very high resistance—up to several megohms—and a very small shunt capacitance—only a few picofarads. The direct input of a scope can be used for many tests as is. It can also be used in any audio-frequency circuit, most video-frequency circuits, for ripple testing, etc.

There are specialized probes to get information concerning special circuits—tuned circuits, very high resistance circuits, etc.—where a minimum of capacitance and a maximum of resistance is needed to keep from upsetting circuit conditions when hooking up test equipment. These special probes have cylindrical housings with test tips of various kinds and are attached to the vertical input of the scope through well-shielded cables to keep them from picking up stray signals—for example, radiation from the horizontal-sweep circuit, hum, etc. The probe bodies are usually shielded as close to the tip as possible for the same reason, and the ground lead is usually kept very short and attached directly to the probe body itself.

The Low-capacitance Probe

Perhaps the low-capacitance probe is the most popular of the specialized probes. Figure 5-5 shows the schematic of a typical unit. Notice that there are actually two capacitors in series in the input. The .05 μF serves mostly as a dc blocking capacitor. Low-capacitance probes are used at high frequencies, and the series reactance of a .05 μF capacitor is so low at such frequencies that you can usually neglect it. The capacitor that counts is the little 9 pF unit shunted across the 1.8 M resistor.

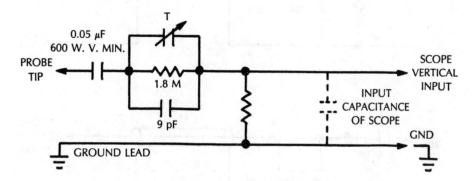

Fig. 5-5. The low-capacitance probe is probably the most popular type of specialized oscilloscope probe.

This capacitor, in series, actually forms an ac voltage divider with the input capacitance of the scope. The size of the capacitor and the value of the cable capacitance are so proportioned that the probe has a 10:1 step-down ratio; it is called a divider probe. Therefore, if you see a 1-volt peak-to-peak deflection on the scope, it means that the actual signal voltage is 10 volts p-p. The little trimmer capacitor shown above the resistor is used to make fine-tuning adjustments of the total probe series capacitance in order to match it exactly to the input capacitance of the scope itself, and to the cable capacitance. Proper matching is needed to obtain the right division ratio and to sharpen the high-frequency response. The resistors used here are all big ones and their exact values are not usually critical.

The Detector Probe

The detector probe seems to be next in popularity. A scope won't show the modulation of a radio-frequency signal until it has gone through the set's detector circuit—the video detector in TV sets, the second detector in radio receivers, etc. If you need to read the signal level, modulation, etc., of a TV or radio signal in the i-f stages, tuner, and other circuits ahead of the detector, you must provide your own detector. Figure 5-6 shows schematics of two detector probes. Of these, the shunt type seems to be used most often, but either one will work. All they do is demodulate the signal so that we can see what's happening to it in the stage under test.

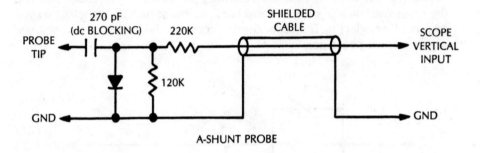

A-SHUNT PROBE

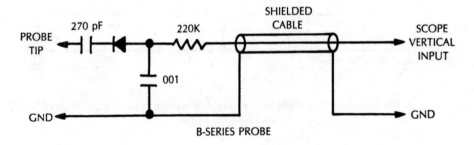

B-SERIES PROBE

Fig. 5-6. Here are two detector probe circuits.

Each probe consists of a crystal-detector diode, with a couple of resistors and capacitors. For safety's sake, a dc blocking capacitor is usually placed between the diode and the probe tip in case you should accidentally touch a high-voltage point in the circuit; 200 to 300 volts dc wouldn't do a 1N64 diode much good.

A detector probe can be used for signal tracing through the video i-f of a TV set if you feed in an AM signal. The special tests you can make with this type of probe are explained as we go along.

There are also voltage-doubling probes with two diodes, but most detector probes are simple half-wave rectifiers. The type of circuit used is usually marked on the probe body.

The Resistive-Isolation Probe

There's one more specialized probe. It's called a resistive-isolation type, and that's just what it is—a resistor, somewhere around 50,000 ohms, mounted in a probe body. This probe is used in some tuner tests, video-amplifier tests, etc., so you get a little more resistance and can isolate the scope's input capacitance from the circuit as much as possible.

The Direct Probe

The direct probe is just what it says: the wire from the vertical input of the scope comes straight through the probe body to the tip. The body of the probe is, or certainly should be, well shielded as close to the tip as possible to hold pickup of stray signals to a minimum.

TESTING FOR FILTER RIPPLE WITH THE SCOPE

One of the best and fastest ways to check the condition of the filter capacitors in all ac power supplies (tube or transistor) is to touch the scope probe to the rectified output—to the B + point that supplies the operating dc voltages to the circuits. The voltage at this point should be a pure dc, which would make a straight line on the scope, but normally there's a small amount of the ac left. This ac is called ripple; it looks like Fig. 5-7.

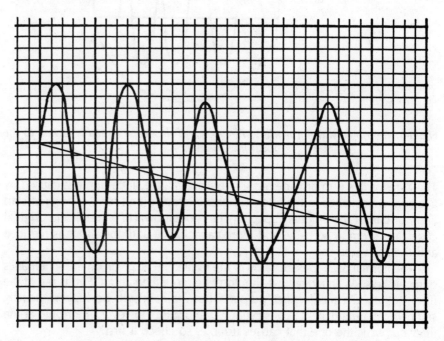

Fig. 5-7. This scope display reveals ripple in the ac signal.

Note the generally triangular shape of the waves, and also note that you see a high peak, then a lower one, then another high one, etc. This unevenness is due to the fact that this is a 120-Hz waveform in a full-wave rectifier circuit; in TV sets, the vertical output circuit takes a heavy pulse of current at a 60-Hz frequency, and this causes the drop in voltage on every second peak of the ripple.

You might have encountered TV's where a dark (or light) bar floats up and down through the picture. This trouble is caused by insufficient filtering in the B+ circuits or by faulty silicon rectifiers. If you check the power-supply ripple, you'll see a big difference between the high and low peaks, and one will usually change phase with respect to the other. This ripple waveform will writhe slowly if you lock the scope sweep to the local line frequency, as illustrated in Fig. 5-8.

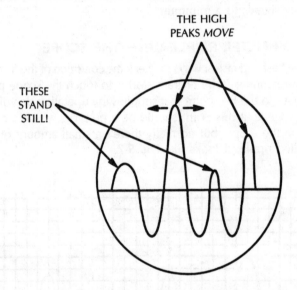

Fig. 5-8. The ripple waveform will writhe slowly if the scope sweep is locked to the local line frequency.

The cure for this is to provide more filter capacitance since ripple is caused by a lack of capacitance in the filter. Either the circuit didn't have enough to start with, or one of the original capacitors has gone down in value. If the original capacitors seem to be okay, add more capacitance until the ripple is reduced and the bar vanishes.

The normal peak-to-peak value of the ripple will be given on the schematics of recent sets. As an average, it shouldn't be more than about 2 volts p-p at the filter output. On the rectifier output (filter *input*) it will be higher—usually somewhere around 10-12 volts p-p, or even more. However, the ripple at the filter *output* is the one that causes trouble, so check it first.

CHECKING CB TRANSMITTER MODULATION WITH THE SCOPE

The oscilloscope will do a good job of measuring the audio modulation of a CB transmitter if we use it correctly. Measure the rf output with and without modulation. However, even the wideband color scopes with bandpass up to 5 MHz won't do too much on a 27-MHz rf signal. You'll get only a line on the scope screen. When you modulate the transmitter, the modulation will appear on the screen. This is what happens: The vertical amplifier in the scope simply won't pass the high-frequency rf carrier, but it will detect the audio modulation and show a goodsized deflection.

To get a true picture of the rf output, feed it directly into the vertical plates of the scope CRT. This gives you an almost unlimited bandpass but no gain, because you're going around the vertical amplifiers of the scope. You'll probably see a line about ½ inch in height at most, for the unmodulated carrier. If you feed an audio signal into the microphone (or whistle into it) you'll see the modulation. The hook-up for this test is illustrated in Fig. 5-9.

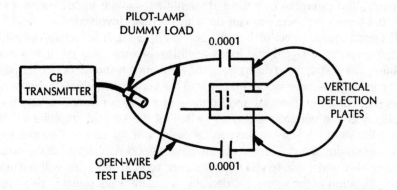

Fig. 5-9. An oscilloscope can be used to check a CB transmitter's modulation with this hook-up.

By estimating the increase in pattern height, you can tell if the transmitter is modulating properly. If it should be overmodulating, the carrier will break up into the characteristic string-of-beads pattern. Check the modulator or rf stages; too much audio or too little rf output have the same pattern and effects.

Another good cross-check for modulation is to use a pilot lamp as a dummy antenna. With the transmitter unmodulated, it should glow a bright yellow; whistling into the microphone should make it glow more brightly.

Some older scopes have provision for connecting directly to the vertical deflection plates by turning a switch on the front marked AMP OUT. Others have terminal boards on the back of the scope case with links that can be opened to give access to the vertical plates.

In either case, use plain, unshielded test leads for this test. Don't use shielded wires: the shunt capacitance of even the lowest-capacitance coax will reduce the signal strength greatly, and there isn't any to spare. Put a small blocking capacitor in series with each lead—a .0001 μF (100 pF). Connect the test leads directly across the dummy-load lamp.

If your scope uses only one vertical-deflection plate with the other tied internally to one of the horizontal-deflection plates, use the free plate as the hot lead and connect the common plate to the CB-set chassis. Keep the test leads well apart, and make sure they aren't moved during the test; movement could change the shunt capacitance and the readings.

SQUARE-WAVE TESTING

A square-wave signal is more than a sine wave with the tops and bottoms clipped, although this is the way many commercial signal generators make them. If you want to get mathematical about it, a true square wave is a signal of periodic recurrence made up of an infinite number of odd harmonics of the fundamental frequency. You can make complicated scientific tests with square waves if you want, but there's one test you can do with no math involved.

If a good square wave, such as the one in Fig. 5-10, is fed into an amplifier and the output is a signal that looks anything like the original, it is a good amplifier. The first thing to do is check the frequency response of our scope by feeding the square-wave signal directly into the vertical amplifier. If your scope shows signs of tilt on the tops and bottoms of the square waves or any other troubles, you can get around this by switching the vertical amplifier off and feeding the signal directly to the vertical-deflection plates. Most scopes have provisions for doing this. You can always have a good-sized signal at the output of an amplifier under test, so you'll get plenty of vertical deflection without using the amplification of the scope. Incidentally, a square-wave signal is also useful for finding horizontal nonlinearity and other troubles in the scope amplifiers.

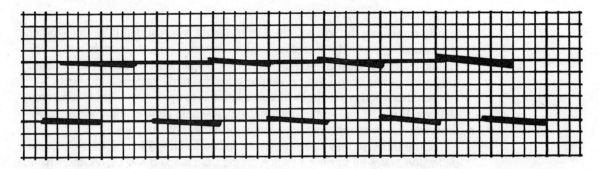

Fig. 5-10. This is a scope display of a good square-wave signal.

In the waveform of Fig. 5-10, all you see are the tops and bottoms of the signal. The vertical lines aren't visible because they rise and fall too fast to register. This is the sign of a good square wave (and also a fairly good scope amplifier, even if it is just a service-type scope). However, even if the vertical lines are visible on your scope, it doesn't matter. What does matter is the distortion that might be added to the square-wave signal on its way through the amplifier being tested; all you need to do is compare the input and output signals, to evaluate the amplifier's performance.

For instance, if the square wave at the output of the amplifier looks like Fig. 5-11, then the amplifier has poor low-frequency response. Poor high-frequency response shows just the opposite reaction; the tops of the waves slant up toward the right.

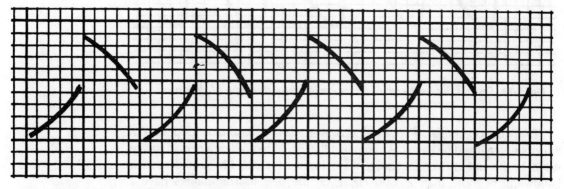

Fig. 5-11. A circuit with poor low-frequency response distorts a square-wave signal like this.

If you feed in a good square wave and get something out like Fig. 5-12, the amplifier is *differentiating* the square pulses (making spikes out of them). This is usually due to a load resistor that has gone down in value or an open capacitor, etc.

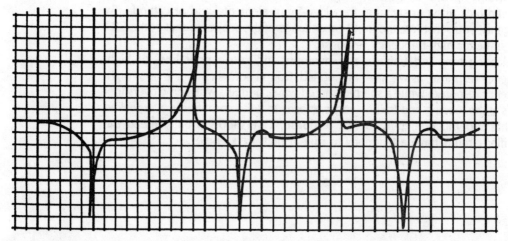

Fig. 5-12. This scope trace reveals that the amplifier being tested is differentiating the square pulses.

The type of distortion seen in Fig. 5-13 could be caused by a leaky coupling capacitor, a load resistor that has fallen off in value, incorrect bias, etc. This basic pattern can also be found in push-pull circuits where the two halves aren't balanced; note that the top halves of the wave differ from the bottom. This always means an imbalance exists somewhere.

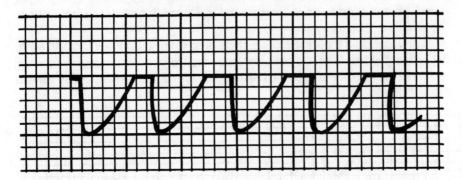

Fig. 5-13. Several different circuit faults could cause this type of distortion.

Don't be too critical. If your output waveform is a reasonable duplicate of the input, take it and be happy. An amplifier that could reproduce a perfect square wave in the output from a square-wave input of 1,000 Hz would need a bandwidth of more than 20,000 Hz. Even the best hi-fi amplifiers have a maximum bandwidth of about 50,000 Hz, and these are the very high-quality, expensive types. Cheaper amplifiers have a maximum bandwidth of about 18,000 Hz, so you'll never find perfect reproduction. In fact, if the output signal is as good as Fig. 5-13, but without the imbalance, take it; it's not so bad.

Square waves can also be used to check video amplifiers. These amplifiers will show better-looking output signals than audio amplifiers because they have bandwidths up to 3 or 4 MHz. To get the best results, pull the socket off the CRT and pick up your signal at the video-input element—grid or cathode. By doing this, you substitute the input capacitance of the scope for that of the CRT and get better high-frequency response.

Slanting tops and bottoms of waveforms indicate a poor low- or high-frequency response, as we have mentioned. Round corners indicate poor high-frequency response, while a sharp-cornered wave, perhaps with some overshoot spikes at the leading and trailing edges, means that the amplifier is overpeaked and has too much high-frequency response. Overshoots usually mean the amplifier circuits have a tendency to ring at high frequencies.

6
Component Tests

This section deals with the problems and solutions in dealing with discrete components.

DETECTING THERMAL DRIFT IN RESISTORS

Thermal drift in resistors is one of the most common and annoying troubles in all kinds of electronic equipment. It causes such symptoms as "it plays for an hour and then acts up," etc. If a carbon resistor changes in value as it heats up, it changes the circuit characteristics.

The time constant will tell you a lot. If the trouble shows up inside of 15 minutes, it is probably due to self-generated heat in a resistor. This is heat generated in the resistor because of the current it is carrying; heat occurs in plate-load circuits, voltage-dropping resistors, etc. If the trouble takes an hour or so to appear, then the resistor is being affected by heat traveling through the metal chassis or from a hot component close by.

The best test for a suspected resistor is to heat it up artificially and watch for the trouble to appear. For example, if we had a long-time-constant sync trouble, we could turn the set on, adjust it for correct operation, and then apply heat to each of the resistors in the sync circuit. Place the tip of a soldering iron on the body of each resistor, hold it there for 45-60 seconds, and watch for the sync trouble on the screen. Normal operating temperature in a TV set is 120-130°F. The tip of a soldering iron runs about 600°F, so don't hold it on the resistor too long—just long enough to get the resistor warmer than normal or too hot to hold a fingertip on it. If the resistor has a tendency toward thermal drift, this test will reveal it.

The high-value (6, 8, 10, and 20 M) resistors found in agc circuits are frequent offenders. You can find the guilty resistor by turning the set on, heating each one up in turn and watching the screen for any sign of the original trouble. There might be more than one defective resistor in a given circuit, so check them all while you're there.

CAPACITANCE TESTING: HOW AND WHAT FOR

There are only two things you need to know about almost any paper, ceramic, or mica capacitor: Is it open, or is it leaky or shorted? But how about measuring the capacitance value? It's seldom necessary, since this information is normally stamped or color coded on the capacitor itself. Capacitors of these types are not likely to shift in value, so the real question boils down to whether it is good or bad.

A capacitor tester is handy for determining whether a capacitor is open. Hook it up and turn the dial rapidly from one end to the other, past the nominal value of the capacitor. If the tuning eye on the tester opens at all, the capacitor is not open.

Make the same check with an ohmmeter for values larger than .01 μF by touching the ohmmeter across the capacitor and watching for the charging kick. (Smaller values give a charging kick too, but it is too small to show on the meter.)

Next, and most important, check the capacitor for leakage. A dead short or high leakage can be caught with the ohmmeter. If the ohmmeter shows any deflection at all on the highest ohms range available, the capacitor is bad. For very critical applications, such as audio coupling capacitors in vacuum-tube amplifiers, you need a test instrument that reads very small leakage. Even a leakage of 100 M is enough to cause trouble in a coupling capacitor.

The fastest test for a possibly open capacitor is to bridge another one across it. If you suspect oscillation is caused by an open bypass capacitor, for example, bridge it. If the oscillation stops, you have found the trouble. If a coupling capacitor is suspected, you can test it in two ways: (1) check for the presence of signal on both the input and output sides of the capacitor; (2) bridge another capacitor across it—if the signal now goes through, the original is open. If a capacitor opens, it has the same effect as taking the capacitor completely out of the circuit. So, replace it by bridging, and see if the trouble stops.

There has been a lot of development in capacitance meters in recent years. Newer devices check for opens and shorts, measure leakage, and directly display the capacitance value. These devices and other new types of test equipment are discussed in Chapter 1.

Electrolytic Capacitors

Electrolytic capacitors, unlike paper capacitors, can change value by drying up. A dry electrolytic is not dry any more than a dry-cell battery is. If the electrolyte evaporates in either one, it stops working. A battery dies, and a capacitor completely opens.

With electrolytics, the best test is again, "How well do they work?" If the service notes specify that a power supply should have 275 volts at the rectifier output and you find only 90 to 100 volts, the input capacitor is very likely open. Bridge it with a good one; if the voltage jumps up to normal, that was the trouble. If the ripple or hum level is given as 0.2 volts p-p at the filter output and your scope shows 10 to 15 volts p-p there, bridge the output capacitor. If the ripple drops to within the proper limits, that capacitor was open.

For bridging purposes, the test capacitor need not be an exact duplicate of the original. It can be much larger or smaller in value, but it must have a working voltage able to stand whatever voltage is present in the circuit. It is not a good idea to bridge electrolytics in transistor circuits with the amplifier on. The charging-current surge of the test capacitor can cause a sharp transient spike in the circuits, and this can puncture transistors. So, to bridge test in transistor sets, turn the set off, clip the test capacitor in place and then turn the set on again. With the instant starting of transistor circuitry, this won't waste any time.

If one unit in a multiple-type electrolytic is found to be bad, replace the whole can. Whatever condition existed inside that can to make one unit go out will eventually cause failure of the rest, because they're all parts of the same assembly. To avoid an almost sure callback, change the whole thing at once.

Finding the Value of an Unknown Capacitor Without a Modern Capacitance Meter

If you don't happen to have a modern capacitance meter handy, there is still a fairly quick way to find the value of an unknown capacitor. This method isn't accurate unless you have a precise ac voltmeter or calibrated scope and an accurate test capacitor. The method is handy for finding the values of those odd mica capacitors that everyone has around and can't read the color code on.

Put the capacitor in series with a known capacitor, as shown in Fig. 6-1, and apply an ac signal voltage across the two. Measure the voltage across the known capacitor, then measure the voltage across the unknown capacitor. The voltage ratio will give you the ratio of the capacitances.

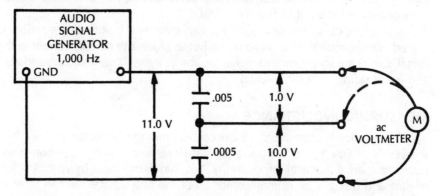

Fig. 6-1. A capacitor's value can be approximated with a test hook-up like this.

What you're doing is putting two reactances in series across an ac-voltage source. The result is a voltage divider that works the same way as two resistors in series across a dc source. Suppose, for instance, that you put 11 volts at 1,000 Hz across a series combination consisting of a .005 μF capacitor and an unknown capacitor. Suppose then you measured 10 volts across the unknown and 1 volt across the known.

Now the figuring: the voltage ratio for the unknown and known capacitors is 10:1, and so is the ratio of reactances. The ratio of *capacitances* is just the opposite, or 1:10. This is because the reactance of a capacitor is inversely related to the capacitance: As capacitance increases, reactance decreases, and vice versa.

In the example given, the unknown capacitor has a value one-tenth as great as the known capacitor, or .0005 µF. Remember, a voltage ratio of 10:1 means a capacitance ratio of 1:10. If you prefer, you can work with reactance values, reading them from a table and then using the table to find the corresponding capacitance values.

Testing Capacitors with a Scope

An oscilloscope is another handy tool for checking capacitor problems. Generally the standard low-capacitance probe is sufficient, although in some cases, a demodulator probe would be useful.

To test the capacitor, adjust the scope for maximum sensitivity on vertical-gain. Then touch the probe tip to the hot side of the capacitor to be tested. Interpretation of the scope trace depends on the capacitor's intended application in the circuit. In most circuits, capacitors are used for bypassing, filtering, or coupling.

A bypass capacitor is always grounded at one end. Unwanted ac components of a signal are shunted to ground. In this test, you want to determine that the ac components have indeed been eliminated from the signal.

A filter capacitor is really very similar to a bypass capacitor, although it is usually somewhat larger in value. A filter capacitor's function is to shunt unwanted signals, such as ac hum, or rf signals, to ground. It is easy to determine if these components are present in the scope trace.

A coupling capacitor, or a blocking capacitor as it is sometimes called, is placed in the circuit to pass an ac signal but to block any dc component in the signal. Check the scope trace to make sure the ac signal is centered around zero and is not riding on a dc voltage.

MEASURING INDUCTANCE

The question is often asked, "How can you measure inductance?" The best answer is, "You can't." Inductance measurements, with common shop equipment, are a practical impossibility. You can figure it out by spending lots of time and doing a lot of mathematics, but the best advice we can offer is, "Don't."

In practical service work, there's seldom a need to read the inductance of a coil or transformer in henrys or millihenrys. You are usually interested in just one thing—continuity. This is a simple ohmmeter test. All service data give the dc resistances of coils, and as long as your resistance reading on an inductor is within 5 percent of the specified value, the inductor is probably all right.

There are only two things that can happen to an inductor: It can open completely (which is fairly easy to find) or it can develop shorted turns. In power transformers, etc., shorted turns give a very definite indication—smoke. In other circuits, such as output transformers, shorted turns cause a drastic loss of output, and you can locate this by elimination tests and power measurements.

Flybacks are a special case; they act more like tuned circuits than ordinary transformers do. They are tested by a special instrument that connects the coil into a circuit and makes it oscillate. The Q of the coil is then read on a meter. However, flybacks can be checked for shorts by reading the cathode current of the horizontal-output tube and then disconnecting all loads, such as the yoke and damper circuits, and reading again. If the cathode current is far above normal, the flyback is internally shorted. When the loads are disconnected, the current should drop to about one-fourth its normal full-load value.

To get an inductance of any particular value, there's one easy way—buy it. You can wind coils all day, trying to get an 8.3-μH choke, but if you call your distributor, he can have a choke coil with exactly 8.3 μH in a few minutes. Coil and transformer makers have a tremendous selection of coils in all conceivable sizes, listed by their inductance, mounting style, etc. By far the easiest way to work with inductors is to go buy an exact duplicate when you need it.

TESTING INTEGRATED CIRCUITS, MODULES, AND PC UNITS

Printed-circuit units are appearing in radios and TV sets in large numbers. These devices range from a simple RC integrator used in vertical sync circuits to the equivalent of a whole amplifier circuit, each in one sealed package. Admittedly, these units are impossible to check in detail because we can't get into them to test individual parts. However, there is a least one reliable check, and that is an output check.

In the integrated circuit, check to see that the proper composite sync signal is present at the input. If the correct sync waveform is not found at the output, the unit is probably bad. This method can be used on any module. There are three things that must be carefully checked before any printed-circuit unit is condemned: (1) the supply voltages and currents; (2) the input signal; and (3) the output signal.

For instance, if the modular circuit is an audio amplifier, it will need a certain amount of dc voltage supply and draw a certain amount of current. With 0.5 volt of audio signal on the input, it has a normal output of 5 volts. If the unit meets these specifications, look elsewhere for the trouble. Don't replace units at random. Make definite tests and be positive before you replace any units.

The scope and signal generator can tell you if a stage is definitely bad by checking input vs output. If it is bad, don't overlook the supply voltage.

A QUICK CHECK FOR MICROPHONES

There's a good, quick check for almost all microphones, especially the common dynamic and crystal types: Make them talk rather than listen. Any microphone can reproduce sound as well as pick it up. For instance, if you have a tape recorder that won't record, the first question is whether the trouble is in the mike or the amplifier. Feed an audio signal into the mike and listen.

You can use an audio-signal generator or any audio signal from a radio or TV set. It takes only a very small signal to make a mike talk. A dynamic microphone is nothing but a specially-built dynamic speaker. Crystal mikes won't talk as loudly as dynamics, but even the variable-reluctance types used in communications work will talk. Incidentally, this is a good quality-check for microphones if the complaint is distortion in the sound output of a PA system or transmitter. By feeding a music or voice signal into the mike and listening to it, you can detect dragging voice coils, buzzes, etc.—defects that would distort the sound pickup. Also, if you happen to have a replacement cartridge for the type of mike you are testing, you can easily make A-B comparison tests of the sound quality of each.

This test can be reversed, too. If you have a complaint of possible mike trouble, hook up a small dynamic speaker to the mike input and talk into it. If the mike input is high impedance, use an output transformer to bring the low voice-coil impedance up enough to work. Actual high-impedance mike transformers will be up around 50,000 ohms, but you can use almost any output transformer. One of the old 25,000-ohm transformers is good, but the test will work with even a 10,000-ohm type. If you're checking for possible mike distortion, hold the mike at least 10 to 12 inches away as you talk. You'll be surprised at the quality of the sound. Transistor radio speakers make good test mikes because of their tiny size.

CHECKING PHONO CARTRIDGES WITH THE SCOPE

When you have low gain in record-playing systems, one of the first things to determine is whether the trouble is in the cartridge or amplifier. The scope is a quick check for the cartridge. With its high-gain vertical amplifier, use it as a sensitive ac voltmeter.

Put a single-tone test record on the turntable, disconnect the cartridge leads at the amplifier (although this isn't really necessary if they're soldered in) and hook the vertical input of the scope to the hot wire, as shown in Fig. 6-2. Set the vertical gain control of the oscilloscope to give about 1 inch deflection for a 1-volt p-p input. Put the stylus on a band of continuous tone—say 400 Hz. For the average crystal cartridge, the output will be from 1-3 volts.

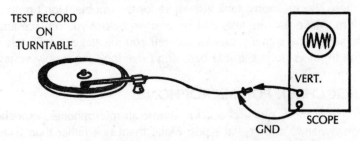

Fig. 6-2. An oscilloscope can be used to check a phono cartridge.

If you want to make a frequency run on the cartridge, you can do it even with a narrow-band scope. Most of these scopes will go up to at least 50 kHz without trouble. You'll need a test record with a band of all frequencies on it, starting at 30 Hz and going to 20 to 30 kHz, at the same output level. Several test records of this type are available. This test can also be used on the whole amplifier.

One valuable application of a scope test is checking stereo cartridges for equality of output in the two channels. Use a test record with a monaural band at about 400 Hz, or a stereo band with equal outputs in the two channels. Several test records have this band for checking speaker phasing, channel balance, etc. Just read the output from each side of the cartridge; the two should be the same.

Cartridge tests can also be made with an ac VTVM. The readings will be the same as with the scope: 1-3 volts p-p. Even an AC-VOLTS scale on the VOM will do, although the meter must have a sensitivity of at least 10,000 ohms per volt on ac. The low input impedance of the VOM reduces the readings. They average from 0.3 to 0.4 volt, where the scope reads 1 to 3 volts. Crystal and ceramic phono cartridges should work into a load impedance of 3 to 4 M.

VIBRATOR TESTING WITH THE SCOPE

There is little equipment around any more that uses vibrators; these devices went out as the transistorized auto-radio came in. However, you might come across one now and then.

The condition of the contact points in the vibrator is a big concern in any vibrator-equipped power supply. With the scope, you can get the answer in a short time. (In fact, the scope is about the only instrument that can give you this information.)

Connect the direct probe across the primary of the transformer. This is usually connected to the small pins of a 4-pin vibrator. If the vibrator is good, you see a nice, clean—though odd-looking—square-wave pattern, as illustrated in Fig. 6-3.

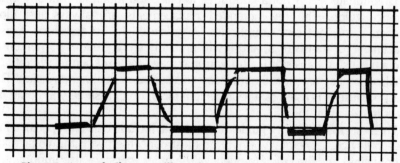

Fig. 6-3. A good vibrator will produce this sort of pattern on a scope.

The top and bottom of the wave represents the contacts, one for each. If either contact is not making a good connection or is bouncing, the pattern will be very ragged and similar to the one shown in Fig. 6-4.

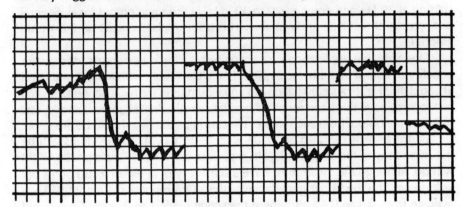

Fig. 6-4. If the scope trace looks like this, one of the vibrator contacts is not making a good connection or is bouncing.

If half of the wave is gone, one contact isn't functioning at all. In all cases, if you see anything in addition to the clean square-wave pattern, put in a new vibrator, because the old one isn't going to last long.

REPLACING TRANSISTORS

It is often difficult to find an exact brand-name replacement for a bad transistor, especially one from an older, discontinued set. Foreign manufactured equipment tends to include a lot of unusually numbered parts.

Many component manufacturers offer a line of general replacement devices, and all good technicians should have as many cross-reference guides as they can find.

Unfortunately, it is not a good idea to place too much faith in any cross-reference guide. The recommended replacements are always just close approximations, not exact duplicates. A guide might list transistor A as a replacement for transistor B, and it will probably work great in 99 percent of the circuits using transistor B. But you may have the exceptional circuit.

Also, don't assume that the replacements can work in either direction. That is, just because A is listed as a replacement for B, don't assume you could use B as a replacement for A.

The best cross-reference guides are ones that include data and specifications for each device. Generally you will be most concerned with the voltage and current ratings and the cutoff frequency. The other specs are also important, of course, but in most applications they offer more room for error. When in doubt, or if you run into problems, use a replacement transistor with slightly higher ratings than the original unit you are replacing.

7
TV Tests

Let's begin by discussing dc current measurements in the sweep and hv circuits. In all tube-type TV sets, both color and black-and-white, all of the power used by the horizontal-sweep and high-voltage circuits is supplied *through* the horizontal output tube. You can tell a lot about the condition of this circuit by reading this current and comparing it to the normal value given on the schematic. Each likely trouble condition is indicated by its effect on the current. A leakage, short, or low-drive condition will make the current go up; an open circuit will make it go down. Disconnect parts to check the effect on the current to get the necessary "split" in the circuit to find a starting point for diagnosis. A typical circuit is shown in Fig. 7-1.

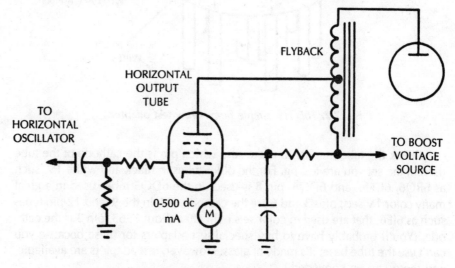

Fig. 7-1. A typical TV high-voltage sweep circuit is shown here.

97

There is an easy way to measure the current of the horizontal output tube. Put the tube on a special test adapter that breaks the cathode circuit. Then, connect a dc milliammeter in series with the cathode and read the total tube current. If you want to, deduct the 12 to 15 mA of screen-grid current to get the exact plate current. However, this isn't usually necessary; use the total current.

Figure 7-2 shows how to build the test adapter. The base from a dead octal tube having all 8 pins is cleaned out. An octal socket is mounted on top. All pins except the cathode are run straight through—1 to 1, 2 to 2, and so on. If you use solid No. 20 wire, it will hold the socket firmly on the base. If it becomes loose, the socket can be cemented in place. Leads for measuring the cathode current are brought out the side as shown—one lead from the socket terminal, the other from the base pin. Test-lead wire, which is very flexible and well insulated, should be used for these. Put pin tips on the ends to match the jacks on your VOM.

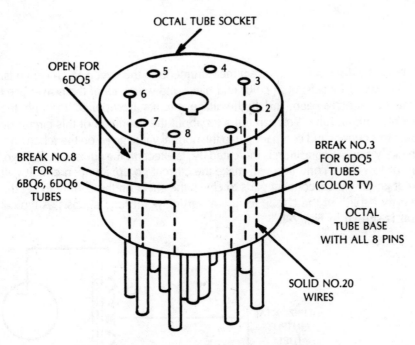

Fig. 7-2. This is a simple home-brew test adapter.

Check the tube manual to make sure which pin is the cathode of the tube used in the set you are testing. For the older tubes in black and white TV, such as 6BQ6, 6CU6, and 6DQ6, pin 8 is used. In the 6DQ5 tubes used in a great many color TV sets, pins 3 and 6 are the cathode, etc. In the 9- and 12-pin types such as 6JE6, that are used in color sets built after about 1965, pin 3 is the cathode. (You'll probably have to buy special test adapters for these because you can't use the tube base; it's made of glass. However, test adapters are available, and these can be converted.)

In many sets, pins 3 and 6 are tied together at the socket; in most cases, pin 3 seems to be the favorite for making the ground connection. The service manuals specify opening this ground lead to hook up the milliammeter, but this means that the chassis must be taken out of the cabinet. With a suitable adapter, you can take the same reading from the top and save lots of time.

Normal current will be specified. After a little practice, you'll learn the average current drain for the popular tube types. For example, 6BQ6 tubes should draw about 95 mA maximum; 6DQ6 tubes can carry up to 120 to 130 mA safely; and 6DQ5's in color sets draw up to 185 to 200 mA. If you find a new tube type, check the tube manual to see what its safe cathode current is, and to learn its pin connections.

CURRENT TEST ADAPTER
FOR 6JE6 AND OTHER NOVAR TUBES

Current-test adapters for Novar tubes are on the market, but you can make one up yourself. Get a 9-pin Novar voltage test adapter, which consists of a base and socket, with a small metal lug brought out from each pin at the top. These lugs are designed for voltage measurements on top of the chassis; the adapter is plugged into the socket, and the tube into the adapter.

To modify such an adapter for breaking the cathode circuit, get a replacement-type Novar socket, preferably of the PC-board type. Set this on top of the adapter socket, as illustrated in Fig. 7-3, and solder the socket pins to the corresponding lugs of the adapter, leaving the cathode pin connections open.

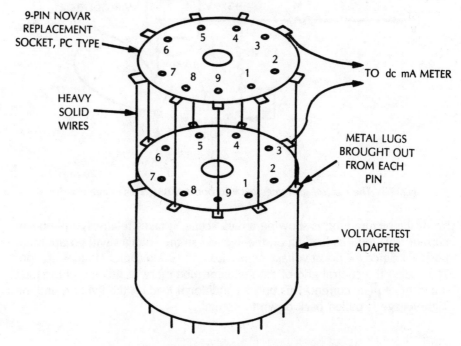

Fig. 7-3. This is an adapter for Novar type tubes.

(The cathode is pin 3 in the 6JE6 tube, among others.) Then connect your meter leads to these two points.

The socket terminals can be connected to the lugs with short pieces of heavy wire if they won't meet well enough to make a good solid joint. The spacing of the adapter in the drawing is exaggerated to show how it's hooked up; the socket should fit down over the adapter much closer than this. Similar adapters can be made for 12-pin tube types and any others.

HIGH-VOLTAGE REGULATOR CURRENT TESTS IN COLOR TV

Many color TV sets use a shunt-regulator circuit to hold the high-voltage output at a constant level. Such circuits are found in both solid-state and tube sets. In a number of older tube sets, a specially designed triode tube, the 6BK4-A, is used. Its plate is connected to the high-voltage line, and its cathode is returned to B+. The control grid goes back to boost through a voltage divider, as you can see in Fig. 7-4.

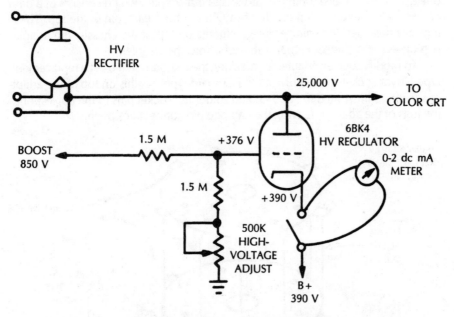

Fig. 7-4. The control grid goes back to boost through a voltage divider.

If the picture tube is showing a dark scene, it takes relatively little beam current. This reduces loading on the flyback, so the voltages will go up (more positive). Since the boost voltage comes from the same source, it goes up, too. This makes the control grid of the regulator tube more positive, and the tube draws more plate current. This puts an additional load on the flyback, and the high voltage is pulled back down to normal.

To adjust this circuit for normal action, we put a 0 to 2 dc milliammeter in the cathode circuit; a special link is provided for this. The high-voltage voltmeter is hooked to the high-voltage connector on the picture tube, with an average scene being shown. The exact value will vary between sets: some call for 24,000 volts, some for 25,000, and so on; check the service data.

Now, check the action of the regulator by turning the brightness control up and down. It should work like this: at maximum brightness, a heavy beam current is going to the CRT, with very little current flowing through the regulator, so regulator current should fall to almost zero. At minimum brightness (dark screen), all of the current will flow through the regulator, because the picture tube is cut off. This current should be between 0.8 and 1.2 mA (800 to 1,200 μA). This value will vary between sets.

If the circuit doesn't work as it should, and the regulator tube draws too much or too little current, check the bias voltage. If any of the resistors in this circuit change value, they will upset the bias and cause the regulator tube to draw too much or too little plate current.

Despite its size, the big regulator tube has a limited plate-current carrying ability; maximum is about 2 mA! So, if the tube is trying to carry too much, it may break down. This can blow the high-voltage rectifier, horizontal output tube, and even damage the flyback, if the fuse doesn't let go quickly enough. Incidentally, the plate of this tube sometimes shows color (glow a dull red) in normal service, when it is carrying maximum current. If the glow is not too bright, it might be okay. To be certain, measure the maximum plate current.

GRID-CONTROLLED HIGH-VOLTAGE REGULATORS

Some color TV sets use a grid-control method of regulating the high-voltage output. The basic principle is the same as that of the shunt regulator, but the circuit is different. Figure 7-5 shows one of these grid-controlled regulators used in several older Motorola color TV chassis.

Control the output of a tube by changing its grid bias. All you need is an indicator of the amount of output. The grid resistor is divided into two sections. On the flyback, a special winding provides a positive-going pulse of 300 volts p-p. This pulse is fed through a blocking capacitor to the anode of a diode rectifier.

To make this rectifier conduct at the right point, a positive bias is placed on its cathode, through a voltage divider from the 280-volt source. The diode will conduct during each pulse, because the 300-volt pulse on the anode is well above the 150 volts on the cathode. When the diode conducts, electrons flow up through the 470K resistor in the grid circuit. This makes the top end of the resistor, the tap in the grid circuit, negative.

If the horizontal output goes up, the pulse goes more positive; that is, higher in amplitude. This puts a higher voltage on the anode of the diode, and it conducts more heavily. This in turn makes more electrons flow up through the 470K resistor, applying a more negative voltage to the control grid of the horizontal output tube. The negative voltage reduces the output, and things go back to normal.

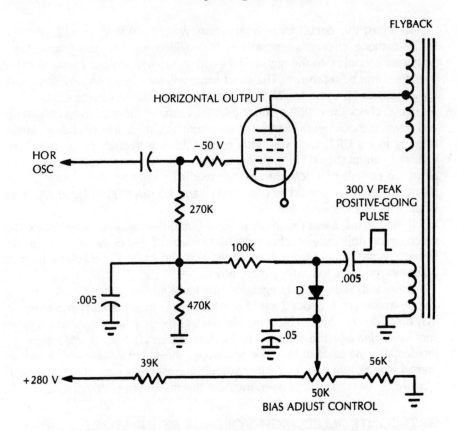

Fig. 7-5. This grid-controlled regulator circuit was used in several early Motorola color TV sets.

If the output drops, the same reaction takes place in the opposite direction. The 300-volt pulse becomes smaller, the diode conducts less, and the smaller current in the 470K resistor allows the grid to go more positive. A positive-going voltage on a control grid makes a tube conduct more; so, the output tube develops more output and the voltages rise until conditions return to normal.

Adjustment of this circuit is simple. A high-voltage voltmeter is hooked to the CRT high-voltage connector, and the bias-adjust control is set to give 26.5 kV. (Check this value on the schematic for the set you are working on, because it might be different; this is the value used in the Motorola TS-912 chassis.)

Servicing should be simple. The diode must be perfect as far as leakage is concerned, and the resistors in the B+ voltage divider must not change in value. Capacitor leakage, especially in the .005 μF coupling capacitor for the high-voltage pulse, would cause a great deal of trouble.

MEASURING VERY HIGH VOLTAGES

In black-and-white TV, it's uncommon to check the high voltage. Use the CRT screen as an indicator. If it's bright enough, assume that the high voltage is okay, and it usually is. In color TV, however, the high voltage must be held within a fairly tight tolerance to avoid purity troubles, etc. So here, you must measure it. This requires a dc voltmeter with a scale of at least 30,000 volts. The easiest way to get such a range is to use an external multiplier probe with our regular dc voltmeter.

Multiplier probes are made with high-voltage insulating housings to keep the user's hands as far from that hot stuff as possible (see figure). The flanges are placed on the base to make the insulation path long. Inside the probe, a special high-voltage resistor—made by depositing a carbon film on a glass cylinder and then coating the whole thing with a high-voltage insulating plastic—is held between spring mountings that keep the contacts tight. The probe housing is sealed to keep any moisture out; moisture might cause flashovers along the multiplier resistor when the probe is in use. A VOM with a high-voltage probe is shown in Fig. 7-6.

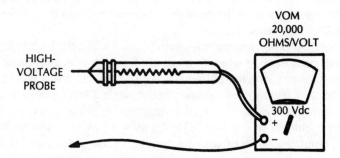

Fig. 7-6. This is a VOM with a high-voltage probe.

If you have a 20,000-ohms-per-volt VOM, you can set it to the 0 to 300-volt dc scale (to make the math simpler). What you need is a meter with a *total* resistance of 30,000 times 20,000 or $(3 \times 10^4)(2 \times 10^4) = 6 \times 10^8$ ohms. Writing out all the zeros, this is 600,000,000 ohms or 600 M. A 20,000-ohms-per-volt meter already has $300 \times 20,000$ or 6 M, on its 0 to 300-volt scale, so subtract this from the total. This leaves a probe resistance of 600 M − 6 M, or 594 M. With a 594-M probe connected, the 0 to 300-volt scale of the meter is multiplied by 100 and reads 30,000 volts full-scale.

A multiplier probe can be used with any meter, provided the correct multiplier resistance is used. Find the resistance of the VOM on the scale you want to multiply (full-scale voltage reading multiplied by the ohms-per-volt rating), then figure the total resistance you'll need to make the meter-probe combination read as high as we want. The meter resistance is subtracted from this total, and the remainder is the resistance we need in the multiplier probe. Probes are available in a variety of resistance values at radio distributor's stores.

For a VTVM or TVM, you have to use a slightly different method to determine the multiplier resistance. When you change ranges on a VOM, you change the meter resistance (0 to 3-volt scale, 60,000 ohms; 0 to 300-volt scale, 6 M; etc.). On VTVMs and TVMs, the total input resistance stays the same for all scales. This resistance could be 11 M, 16 M, or whatever the instrument designer decided to use.

To figure the probe resistance, use the same *principle* as before, but you get a different result. Say you have a 0 to 500-volt scale and you want to make it read 0 to 50,000 volts. This calls for multiplication—multiply by 100. First find the input impedance of the VTVM. Say that this is 11 M, a common value. Of this resistance, 1 M is in the regular DC-VOLTS probe, so the resistance of the internal voltage divider is 10 M. In most VTVMs, the regular volts probe is disconnected and the high-voltage multiplier probe plugged in its place. So, the meter has a resistance of 10 M. To multiply readings by 100, multiply the input resistance by that figure, obtaining 1,000 M as the required meter-plus-probe resistance. Subtracting the 10 M we already have, the multiplier resistor in the probe will be 990 M.

Here's the difference between the VOM and the VTVM: since the input resistance on a VTVM never changes, multiply all of the dc voltage ranges by the same figure when you hook up the high-voltage probe. Even the 0 to 3-volt range becomes 0 to 300 volts, etc.

If the regular probe resistor stays in the circuit when the high-voltage probe is attached, consider that when figuring out the high-voltage resistor value. In the case just figured, if meter resistance was 11 M, the multiplier would be 1,100 M − 11 M, or 1,089 M. One popular meter uses a 22 M input resistance with a 7 M resistor in DC-VOLTS probe. By removing this probe you have 15 M left; so a 100:1 multiplier would require 1,500 M − 15 M, or 1,485 M. You'll find the exact input resistance of your own meter in the instruction book.

MEASURING FOCUS VOLTAGES IN COLOR TV

In color TV, you often need to check the focus voltage. The best way is to take this reading right at the base of the color TV CRT, to make sure that this important voltage is getting to where it is used. Also, check the action of the focus control to see if it gives the proper amount of variation. This variation is usually 4,000 to 5,500 volts dc. This voltage comes from the flyback, through a special rectifier—sometimes a small tube, sometimes a special high-voltage silicon rectifier.

In the original 21-inch color tubes and in many later ones, the focus electrode is pin 9 (check the schematic to make sure). You usually can tell just by looking at the set—many CRT sockets leave a blank space on either side of pin 9 to provide more insulation for the high focus voltage. The focus voltage is measured with a high-voltage probe, as illustrated in Fig. 7-7.

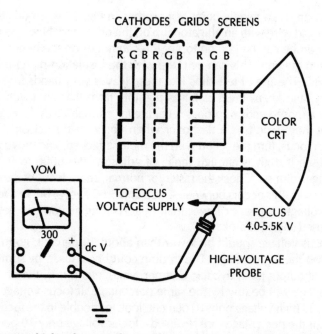

Fig. 7-7. A high-voltage probe can be used to measure the focus voltage.

There's an easy way to make a focus voltage measurement. You need to make contact on the base pin of the CRT itself. This is hard to do if the CRT socket is pushed tightly on (as it should be). In some sets, you can pull the socket back just a little and get at pin 9 with a thin test prod. However, some of the other pins might not make good contact if this is done. The answer is to make up a gadget that will let you get at this pin without disturbing any of the others.

One such device is a small clip made out of spring wire, as in Fig. 7-8A. This is slipped over pin 9, and the socket is then pushed back tightly. The meter prod is then touched to the exposed end. (Just make sure that the meter prod is the *only* thing that touches; 5,000 volts can bite!)

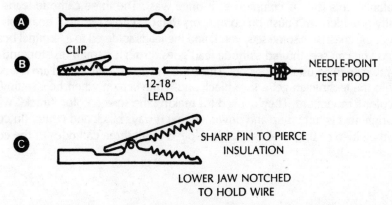

Fig. 7-8. A) Home-made wire for single-pin access. B) Needle-point test prod. C) Clip with insulation-piercing pin.

Or, you can use a needle-point test prod as in Fig. 7-8B. Fit this to a short piece of test-lead wire with an alligator clip on the other end. Now the regulator high-voltage probe can be laid on top of the cabinet or on the bench, with the clip hooked to the end. This can be done with the insulation-piercing clip, too as in Fig. 7-8C. The use of test clips like these leaves your hands free to adjust controls, etc., and keeps them as far as possible from that "hot stuff." Just be sure that the exposed ends of probes, etc., are far enough away from grounded parts so that there won't be a flashover when you turn the set on.

To check focus, turn the set on with the meter attached, and move the focus control and see if it gives enough range of adjustment—4,000 to 5,500 volts for the average color set. Check the raster for normal, sharp focus of the scanning lines. In most sets, this occurs between 4,600 and 5,000 volts; but in a few cases, you'll find voltages higher or lower than this. As long as you can get a good, sharply focused raster, it's okay.

If the focus voltage should be lower than about 10 percent, go to a higher scale and read the high voltage. The trouble could be caused by something in the horizontal output tube, flyback, damper, yoke, etc. If it is, then *both* focus and high voltage will be low by the same percentage. If focus voltage is down 40 percent, and high voltage only 10 percent, look for trouble in the focus-rectifier circuits, etc. If the percentages are reversed—focus voltage only 10 percent low and high voltage 40 percent low—check the high-voltage rectifier, voltage regulator, and the circuits that affect only the high voltage output.

CHECKING COLOR PICTURE TUBES WITH A VOM

Your VOM can be a pretty good picture-tube tester, especially with color CRT's. One of the best ways to measure the quality of any picture tube is to read its beam current, which is the cathode current. It runs 300 to 400 μA maximum in most black-and-white tubes, and about the same for all three guns (total) in color picture tubes.

Many color TV sets have provisions for unplugging each cathode lead of the picture tube, as shown in Fig. 7-9. This is done so that the technician can adjust the beam currents of the three guns, if there happens to be a phosphor imbalance (this isn't as common as it once was). The three cathode leads are usually provided with push-on connectors that fit onto a terminal board on the back of the chassis. In some sets, you'll find the leads soldered to a terminal board.

As you can see, the red cathode lead goes through a small resistor, and the other two go to adjustable resistors. These are the drive controls and are intended to help the technician get a true black-and-white screen when he's setting up the color temperature. They're used for making the screen color "track" when the brightness is turned up and down, so that it stays black and white. (In color sets, the video or brightness signal is applied to all three cathodes of the color picture tube.)

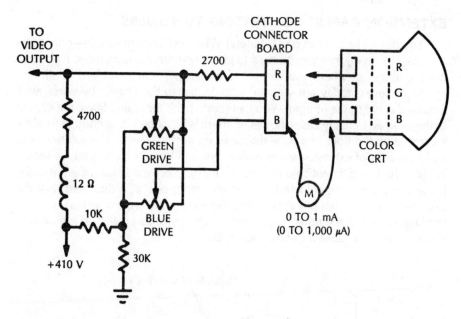

Fig. 7-9. Many color TV sets have provisions for unplugging each cathode lead of the picture tube.

To check beam current, unhook the cathode of the gun you suspect. Connect a 0 to 1-mA dc meter in series with it. Turn the set on and adjust the screen for average brightness; you should read about 100 microamperes (0.1 mA) on each gun, within about 10 percent. If one of the guns has an exhausted cathode, its beam current will be very low, and nothing you can do—such as turning up that screen voltage, etc.—can bring it up to where the other guns are running. Use the currents of the other two guns as a standard; all three should match.

If one gun has a heater-cathode short, you'll probably see all green or all red or whichever one is faulty. Checking the cathode current of the defective gun will probably show you that it is running up to 1 or 1.5 mA instead of the normal 100 μA. The brightness controls will have no effect. Measuring the cathode-grid voltage will probably show you that this gun has zero bias. (Note: If you read 175 volts on the cathode and 175 volts on the grid, that gun has zero bias. The difference voltage between grid and cathode is what you read. Example: cathode has 150 volts and grid has 125 volts. That means the grid has a −25-volt bias on it.)

If this test shows that a gun is bad, verify it by checking the color CRT on a good CRT tester. If this gives you the same answer, then the gun is probably bad. Considering how expensive color picture tubes are, never take the word of only one test before deciding that a color tube is bad. The trouble might be in the operating voltages; check all dropping resistors, supply voltages, etc., before making up your mind that it's the tube. As said before, the voltages and operation of the other two guns are handy as a standard, because all are fed from the same supply in most sets.

EXTENSION CABLES FOR TESTING TV TUNERS

You don't like to work on TV tuners? Why not? They're simple—only a few circuits. However, they can be hard to get to without the right tools. If you can bring a tuner out where you can get to things, servicing isn't bad at all.

Many tuners today are separate—connected to the chassis by wires, with a coaxial cable to the i-f input. You can remove the tuner and leave the chassis in the cabinet. In many cases, however, these wires aren't long enough to leave a console on the floor and have the tuner up on the bench. The solution is to make up a set of extension wires with a terminal board on one end, as shown in Fig. 7-10. Use different colored wires so that you can easily keep them straight. If the tuner wires are soldered, disconnect them one at a time and tack the extension wires in their place, fastening the ends of the original wires under the terminal screws. If the tuner has a plug-in cable, you can make up a plug-and-socket extension cable for about 50 cents.

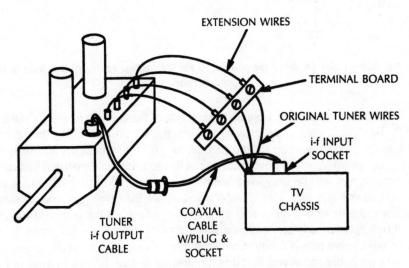

Fig. 7-10. You can easily make an extension cable to test TV tuners.

The i-f output cable from the tuner is often a plug-in type, with plugs like those used on phonographs. Make up an extension for this, too, with a lug and socket. Any kind of small coax will do, because you're only adding a very few picofarads of shunt capacitance. The picture and sound come through in surprisingly good shape.

You can make a rack to hold the tuner while it's on the bench, or just block it up with a couple of empty boxes. Now you can test alignment, make voltage measurements, and make gain checks, etc. a lot more easily than before. Except for the i-f output, all of the alignment adjustments of the tuner can be set up while the tuner is on the extension cables. The i-f output can be adjusted easily after the tuner is back on the original cable.

READING DC VOLTAGES IN CASCODE RF STAGES

Many tube-type TV sets feature cascode amplifier stages, using such tubes as the 6BZ7, 6BQ7, 4BS8, and so on. It is basically a stacked-stage circuit. The plate of the input triode is connected to the cathode of the output section, and has about half of the applied voltage on it. If you take the tube out of the socket, this voltage can't be measured because it won't even be present unless the tube is in place.

Test adapters are available at all radio supply stores. The typical adapter consists of a plug with a tube socket on top, with each pin of the plug wired to the corresponding socket terminal. Around the top of the adapter are small, numbered lugs, each connected to one socket terminal. The adapter is plugged into the tube socket of the tuner, and the tube is plugged into the adapter socket. Now you can touch a VTVM probe to any lug and read the actual voltage with the tube in operation. This type of test adapter is illustrated in Fig. 7-11.

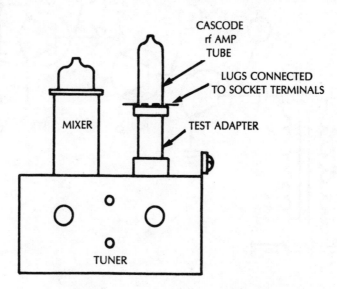

Fig. 7-11. This is a common test adapter for tubes.

A test adapter can be used in any stage, of course, but it is essential in cascodes. Check the maximum plate voltage, which will be on the output-section plate. Then measure the voltages on the input-section plate and the output-section cathode. This should be almost exactly half the supply voltage on the output plate. Also, while you're there, measure the dc voltage on the grid of the output-section triode. This should be only slightly negative, relative to the plate-cathode combination; in one actual tuner, this grid reads 120 volts to ground and the plate-cathode reads 125, giving a −5-volt bias. If you find an imbalance, replace the tube first, then check coils and resistors.

A QUICK TUNER GAIN CHECK

Here's a test that will help you isolate that troublesome stage in a tuner or clear a stage of suspicion. You can use a signal generator, but your bench TV antenna works just as well. If there is too much snow in the picture, it is usually a case of tuner trouble. The only question is, "Which stage?" By using this signal-tracing method, you can pin it down in a hurry.

Figure 7-12 shows the signal path of a fairly typical cascode tuner. Starting at point A, hook the antenna to the input terminals. If the picture is very bad (snowy) and the station is known to give a good clear picture on a normal set, there's trouble. Using the antenna as a signal source, make gain checks through the tuner to find out which stage isn't doing its job.

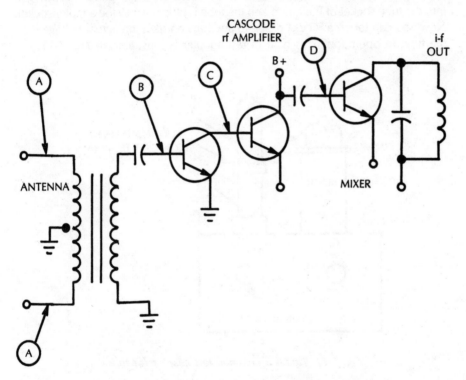

Fig. 7-12. This is the signal path through a typical cascode tuner.

The antenna has a 300-ohm balanced impedance. The output impedance of the balun coil is 75 ohms, unbalanced. However, by grounding one side of the lead-in, use the other side and get a satisfactory match for what you want. Put a capacitor of about 100 pF in series with the other side of the lead-in, just in case dc blocking is needed to keep from shorting out bias or dc voltages. In some tube circuits, this capacitor might be left out, but it is essential for transistor tuners. If the capacitor is not used in a solid-state tuner, the bias on the rf transistors could have damaging results. To be on the safe side, always use a small dc-blocking capacitor in this type of test.

First, touch the free end of the capacitor to point B, the output of the balun. If the picture clears up, the balun coil is bad. If there is no improvement, keep on going.

If you get a better picture by touching down at point C, the first half of the cascode circuit is bad. Change the tube and check the dc voltages. If this shows no improvement, then go on to point D, the mixer grid. If you get a better signal by touching here, then the rf amplifier is faulty.

MEASURING COLOR TV HIGH-VOLTAGE-REGULATOR CURRENT WITH A SERIES RESISTOR

The original method of reading the high-voltage-regulator cathode current in color TV sets was by opening the cathode circuit and inserting a 0 to 1 dc milliammeter. This method is still used in many sets. Some makes, however, have a 1,000-ohm 5 percent resistor in series with the cathode, as shown in Fig. 7-13.

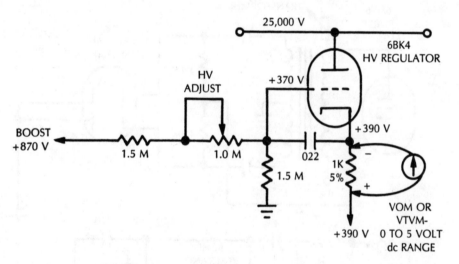

Fig. 7-13. In some sets, a 1K resistor is in series with the cathode.

This resistor saves unsoldering the cathode lead. The meter, which can be either a VOM or VTVM, is hooked directly across the 1,000-ohm resistor; it reads the voltage drop resulting from the cathode current flowing through it. By making the resistor an even 1,000 ohms, there's a 1-volt drop for every milliampere of current that flows. This resistor is an automatic Ohm's law calculator; E = IR, so that 1,000 × .001 = 1.

When the meter is hooked up to this resistor, look out for hot-case effects if your particular meter happens to have its negative return made directly to the case. The negative lead of the meter here is 390 volts *above ground* potential; if you accidentally touched both the chassis and the metal case, you'd be in for a shock—quite literally.

The average current in high-voltage-regulator tubes is small, so your reading will be in the neighborhood of 1 volt. Pick the meter range so that this is as close to center-scale as possible.

FIELD ADJUSTMENT OF COLOR AFPC

Color automatic frequency and phase control (afpc) is a very complicated-sounding circuit, isn't it? Yet, it's no more complicated than the horizontal afc circuits that we've been adjusting for years. It better not be—it's the same circuit. You can make accurate field adjustments on this circuit if you have to, even without instruments. It depends on knowing what the circuit is supposed to do and where the test points are. The circuit shown in Fig. 7-14 was used in many RCA tube chassis and is simple to adjust. More recent, solid-state circuits can usually be adjusted in a similar manner.

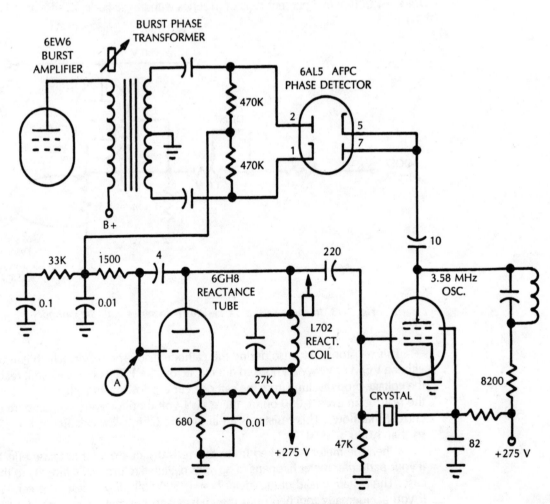

Fig. 7-14. This is a typical AFPC circuit.

The first thing to do is get the set working. Then tune in a color program and see if the color is correct, using human faces as a guide. If not, set the hue control in the center of its range and the color control for a medium color picture. Now ground the grid of the reactance tube (point A in the figure). This takes off all color-sync control, so the colors will promptly fall completely out of sync. You'll see rainbows of color sweep over the picture.

Adjust the reactance-tube plate coil (L702) for a color zero beat. When you get close to this point, the color bars will become very broad; watch for the point where you get only one or two sets of colors floating across the screen. These will react exactly like the pattern you get with an off-frequency horizontal oscillator; when they're far out of sync, you'll see many fine bars of color, slanting almost straight across the screen. As you get close to the right point, you'll have fewer bars, and they'll be closer to straight up. When you hit exactly the right place, you'll see only three colors—red, blue, and green. These will suddenly snap into place and the picture will have color again. (If you keep on turning the coil core, the color will fall out of sync and slant the other way.)

When you reach the exact adjustment, you'll have a zero beat between the color and the set's 3.58-MHz oscillator, and the colors will be correct. Leave the adjustment right here and remove the short from the reactance-tube grid. Now the color should snap in and lock firmly.

If the color suddenly goes out when you put the reactance tube back in the circuit, the oscillator is okay and the reactance-tube circuit has trouble! Check it and find what's wrong.

CHECKING FOR OPERATION OF
THE 3.58-MHZ COLOR OSCILLATOR

For a loss of color or color sync, one of the first things to find out is if the 3.58-MHz oscillator is running. In the circuit used for many years in RCA sets, this is not difficult. The burst and 3.58-MHz signals are fed into a phase detector, as shown in the figure. This stage is exactly like the common horizontal-afc phase detector or an FM ratio-detector circuit.

To check the oscillator activity, measure the dc voltage developed across one of the phase-detector diodes. As you can see, in normal operation the two develop equal voltages of opposite polarity, so you can read either one.

To eliminate the influence of any incoming signal, short the input of the burst-amplifier to ground. If the voltage isn't high enough, adjust the bottom core of the 3.58-MHz oscillator transformer for a peak reading on the voltmeter. Then, the short on the burst-amplifier input is removed, and the burst-phase transformer is adjusted for a maximum voltage reading. If there is trouble in any of this circuitry, these voltage readings will promptly tell you.

USING A PILOT LAMP TO ADJUST
A HORIZONTAL EFFICIENCY CONTROL

The horizontal-linearity control, or horizontal-efficiency control in color sets, should be adjusted for the point of minimum current in the output tube, just as in tuning a radio-transmitter plate tank for a dip. Since both circuits are resonant, they are actually the same type of adjustment. We can connect a milliammeter into the circuit for the most precise adjustment, but there is another way to adjust the circuit, if necessary.

Make up a gadget with a pilot-light socket, the plate cap from a dead tube, and a plate-cap clip, as shown. For color sets, use a 250-mA pilot lamp; for black-and-white sets where currents run about 100 mA or less, use a No. 47 lamp. Pull the plate cap of the horizontal output tube and hook in the adapter, as shown in Fig. 7-15.

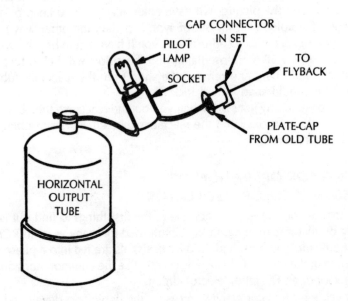

Fig. 7-15. A simple adapter built around a pilot lamp can be used to adjust a horizontal efficiency control.

Turn the set on and note the amount of glow in the pilot lamp. Set the controls (brightness, etc.) as specified in the service data, and tune the horizontal-efficiency or horizontal-linearity control for a dip. When you hit the right point, you'll see a decided drop in the brightness of the pilot lamp. That's it.

The lamp can also be used to check for intermittent shorts or overloads in this circuit. For example, the normal current here in a color TV set is about 200 mA. If a 250-mA pilot lamp burns a bright blue-white, look out! That means too much current.

SCOPE TESTS FOR PRESENCE OF
HIGH VOLTAGE, HORIZONTAL SWEEP, ETC.

The TV screen is dark; is there any high voltage? Is the horizontal output tube working? Is the horizontal oscillator working? We can answer these questions very quickly with a scope, and get a handle on the problem in a very short time.

Hold the scope probe near the plate lead of the horizontal output tube. Don't touch the plate cap; the pulse voltages there will break down the input capacitors of the scope and can cause other damage. You don't need to touch it, anyway. You'll get plenty of pattern height from the tremendous pulse voltages, as you can see in Fig. 7-16.

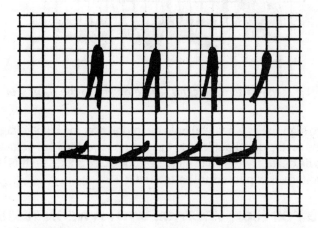

Fig. 7-16. Even without touching the plate lead of the horizontal output tube, a scope will still display plenty of pattern height.

Check the horizontal oscillator frequency at this point by using the frequency-set method outlined in an earlier test. If you see too many or too few cycles, the horizontal oscillator may be so far off frequency that the output stage will not work.

To check the high-voltage rectifier tube, hold the probe near its plate lead (or even near the bulb). You'll see the same pattern, but with much higher spikes since the pulse voltage is very high at this point if the output tube and flyback are okay. Holding the probe near the high-voltage rectifier output lead to the picture tube will also show you the same spikes if the filtering is okay. However, if you find a pattern with small waves on the horizontal parts, as in Fig. 7-17, this indicates a severe ringing in the flyback or yoke. In a few cases, an open high-voltage filter capacitor can cause this same symptom. Also, trouble in the horizontal-linearity coil or capacitors, or a bad balancing capacitor in the horizontal yoke can make this kind of pattern.

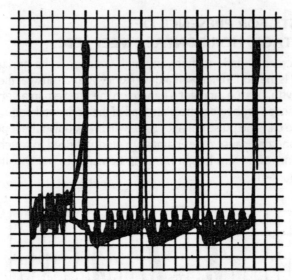

Fig. 7-17. This scope pattern indicates severe ringing in the flyback or yoke.

If you see high spikes when you hold the probe near the high-voltage rectifier plate but none on the dc output lead to the picture tube, the high-voltage rectifier is very apt to be dead. There will usually be good-sized spikes on this lead, even though its voltage is supposed to be filtered.

CHECKING FOR HIGH VOLTAGE WITH A NEON TESTER

A neon lamp will glow in a strong electric field. Such a field exists around the flyback and the plate leads of both the horizontal output tube and the high-voltage rectifier in all TV sets. Special high-voltage testers are made in the form of a rod of insulating material with a neon lamp in a clear plastic housing on one end. If you hold one of these lamps near the plate lead of the high-voltage rectifier and it glows brightly, you know that the horizontal output tube and flyback are working. There is definitely a good deal of energy around there. If there is no dc high voltage, the high-voltage rectifier tube is probably bad.

A neon lamp will also serve as an indicator of the presence of energy around the plate lead of the horizontal output tube, although the lamp won't glow as brightly as it will when near the high-voltage rectifier plate; the rf field here isn't quite as strong, but there will be a definite glow.

SETTING A HORIZONTAL OSCILLATOR ON
FREQUENCY BY COMPARING WITH VIDEO

You'll find many TV sets in which the horizontal oscillator is obviously off frequency. In some, the oscillator can be thrown so far off by misadjusted controls that you lose the raster, boost voltage, etc. (Kids and unqualified technicians are sometimes to blame.) You need a quick way to check the frequency of the oscillator.

Once again, we use a comparison method. The standard is the horizontal-sync pulses of the video signal. Hook up the scope with a low-capacitance probe and pick up a signal at the video-amplifier plate. Adjust the scope sweep to about 7,875 Hz until you see two horizontal sync pulses on the screen, as illustrated in Fig. 7-18. Set the frequency control and sync lock of the scope to hold them as steady as possible.

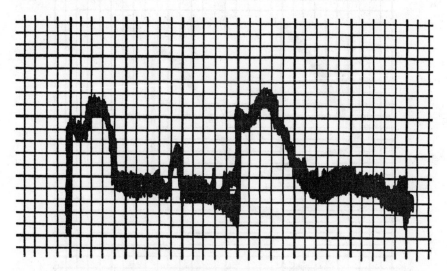

Fig. 7-18. To set the horizontal oscillator frequency, first adjust the scope to see two horizontal sync pulses on the screen.

Now, without touching the sweep controls of the scope, move the probe over to the horizontal-oscillator circuit. The grid of the horizontal output tube is a good place to pick up a signal. You can adjust the vertical-gain controls to keep the pattern on the screen, if necessary. Without touching the scope sweep controls, adjust the horizontal-hold control, coils, etc., until you get two cycles of the horizontal frequency on the scope screen, as shown in Fig. 7-19.

If misadjustment was the only trouble, your raster will come back and you can use the TV screen to make the final adjustment.

TESTING FREQUENCY IN VERTICAL-OSCILLATOR CIRCUITS

If you need to test the frequency of the vertical oscillator and see no picture for some reason (no high voltage, no picture tube, etc.) you can make the test with a scope. Compare the oscillator signal with the video signal used to test horizontal-oscillator frequency. Set the scope frequency to show two or three cycles of video signal, near 30 Hz. You can now see the vertical sync pulses, as in Fig. 7-20. Here again, the pattern is badly blurred, but all you need are the two vertical sync pulses—the bright pips at the top of the waveform at the right and center. (You can spread this pattern and make it easier to see, but we have deliberately used this blurred pattern to illustrate how easy it is to identify the sync pulses.) Move the scope probe to the vertical-oscillator circuit, and adjust the vertical-hold control to show two cycles.

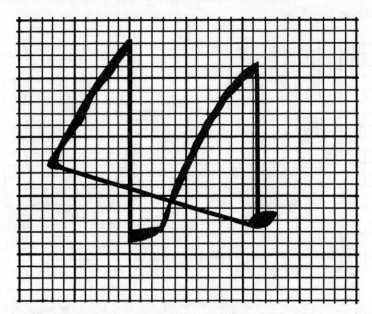

Fig. 7-19. Without touching the scope's sweep controls, make the necessary adjustments to get two cycles of the horizontal frequency on the screen.

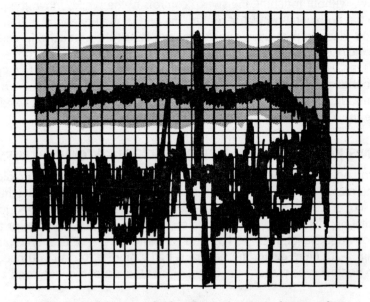

Fig. 7-20. This scope display shows the vertical sync pulses.

There is another way. Set the scope for line sweep, which is a 60-Hz sinusoidal sweep instead of the sawtooth. Now feed in a pulse from the vertical-oscillator circuit. By juggling the vertical- and horizontal-gain controls, you can make the trace an oval, or even a circle. This is illustrated in Fig. 7-21. The notch is the spike from the vertical oscillator. If you see only one notch, and it's standing fairly still on the circle, the vertical oscillator is running at 60 Hz. If the frequency is off 1 Hz, the notch will go around the circle once each second.

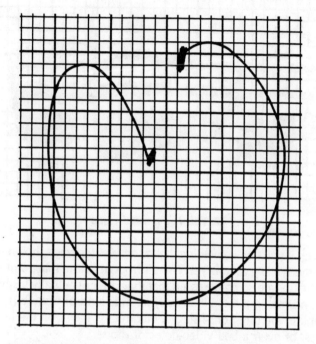

Fig. 7-21. By juggling the vertical- and horizontal-gain controls, the trace can be made an oval, or even a circle.

This procedure is often needed in transistor TV sets that have several ailments at the same time. For example, in one make, if the vertical circuits are out of adjustment, the vertical-output stage will draw such a heavy current that it will kick the circuit breaker, causing you to waste time looking for nonexistent shorts. Setting the vertical oscillator on frequency will eliminate this, even if you have no high voltage and can't see a raster.

FINDING THE CAUSE OF SYNC CLIPPING WITH THE SCOPE

If you find a TV with a case of sync clipping, or sync trouble of any kind, there are several possible causes. The only quick way to find the trouble is to separate the various possibilities with a scope.

To begin with, check the video signal at the sync takeoff point. You can use the direct probe on the scope since this is a high-level signal. You should see a normal video signal, as shown in Fig. 7-22, the sync tips form 25 percent of the total height of the pattern, and the video forms the other 75 percent. If you find a pattern like this, then the video detector and video output are okay; the trouble is actually in the sync-separator circuits.

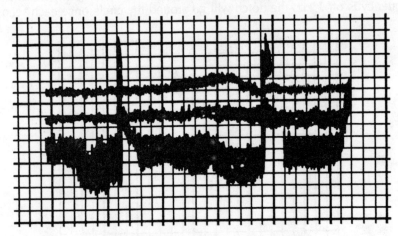

Fig. 7-22. This is a normal video signal.

However, if you find something that resembles the trace shown in Fig. 7-23, don't go to the sync separators yet. There's trouble before that point. Notice that here the sync tips are so compressed that they are not actually visible. The video is about normal, but there is little sync. This kind of trouble is usually due to a bad tube in the tuner or i-f stages, incorrect voltages on the i-f's, incorrect agc voltages, or even a bad video-detector diode.

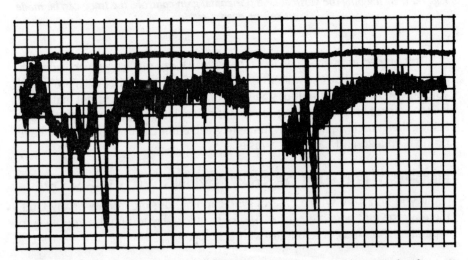

Fig. 7-23. This trace indicates problems ahead of the sync separator circuits.

The symptom in the signal shown in Fig. 7-24 is the exact opposite. It is referred to as white compression. This signal is also a clipped waveform, but notice that the sync tips are normal in height. It's the *video* that is being clipped. The result is a nice, clean raster (no snow) with the picture signals visible only as vague outlines on a very bright background. In bad cases, you usually have to turn the brightness far above normal to see anything.

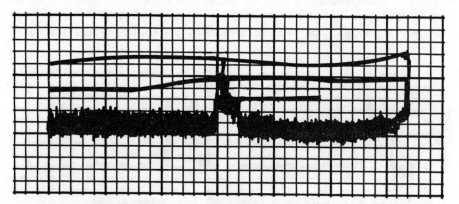

Fig. 7-24. This trace indicates a problem known as white compression.

The most common cause of this symptom is a video output tube that is very weak, gassy, or has heavy grid emission. In some cases, a bad video-detector diode will do it; you'll have to go by the scope patterns on the input and output of the video amplifier. If the signal is clipped on the video-amplifier grid, the trouble is being fed into the tube and does not originate there. If you do suspect the video amplifier, replace the tube first, then check for correct operating voltages, especially grid bias.

White compression is sometimes confused with a weak picture tube because of the loss of contrast. However, there is one basic difference: if white compression is the trouble, the picture tube will still be able to make a very bright, well-focused raster. If the CRT is weak, you'll have very low brightness, and probably loss of focus, especially in highlights. With white compression, highlights smear out but are very bright. If you get about 50 volts p-p of good clean video at the grid or cathode of the CRT but still have a poor picture, then the picture tube should be suspected.

FINDING SYNC TROUBLES WITH THE SCOPE

When we encounter sync troubles, there's only one instrument that will get in there and tell us anything useful—the scope. All sync trouble can be found in the least time by following the sync signal from the point of origin through the various sync-separator and amplifier stages until we find the point where it disappears. Voltage and resistance checks will then pinpoint the cause of the trouble.

Because most of these are high-impedance circuits, we use the low-capacitance probe to keep from disturbing them too much. Start at the sync takeoff point which is usually in the plate circuit of the video-output stage. Make sure the composite video signal is there—with plenty of sync—before you go any further. The signal should look like the one shown in Fig. 7-25. You might see signs of video compression, and in fact, our sample signal in the figure shows a sync-to-video ratio that is closer to 50:50 than to the 25:75 ratio we look for in the full video signal. However, this waveform was taken off at a point lower in the video-output circuit than the full video signal applied to the picture tube.

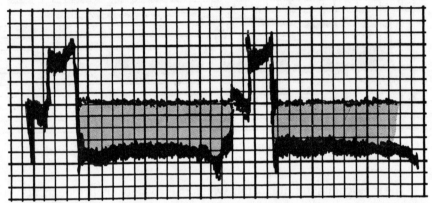

Fig. 7-25. The composite video signal should contain strong sync pulses.

The correct scope patterns for various points in the sync circuit are usually shown in the service data or on the schematic diagram. For instance, the stripped signal might be similar to Fig. 7-26 after the video has been clipped. This is the type of pattern seen at the input to the vertical integrator in many sets. The sharp spikes are vertical sync; the rest are horizontal sync pulses.

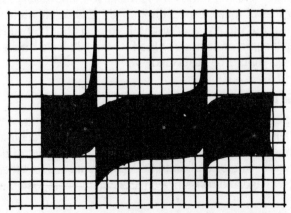

Fig. 7-26. After the video has been clipped, the stripped signal might look like this.

Vertical-oscillator circuits depend mainly on the amplitude of the sync for proper operation. Horizontal sync works mostly on phase and will lock in on a weak signal far longer than will the vertical. So, any weak-sync condition will show up as vertical-roll troubles first.

To check for sync amplitude, hook the scope to the vertical oscillator circuit. Figure 7-27 shows a typical pattern on an oscillator grid, and Fig. 7-28 shows one as seen on a plate. By turning the vertical-hold control so that the blanking bar rolls slowly down, the vertical-sync pulses appear as pips on the waveforms. Although some vertical oscillators lock satisfactorily on less sync than this, it's always nice to have a sync pulse of this amplitude for good, tight-locking action.

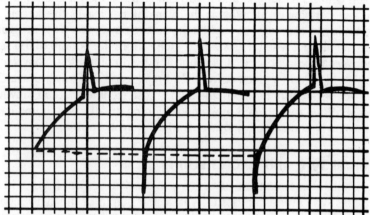

Fig. 7-27. This pattern should be found at the grid (or base) of a vertical oscillator circuit.

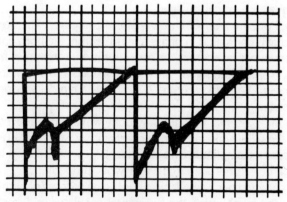

Fig. 7-28. This pattern should be found at the plate (or collector) of a vertical oscillator circuit.

Ordinarily, you just look at this waveform to get a good idea of the sync amplitude (in proportion to the amplitude of the full waveform). If you want to, read the sync-pulse amplitude by itself. Kill the vertical oscillator, and take a reading of the sync at either the grid or the plate of the vertical oscillator—depending on where the sync-injection point is in this circuit. In all cases, it will be found at the output of the vertical integrator, coming from the last sync-separator or sync-amplifier stage. The amplitude of sync for the set getting tested is usually given on the schematic. In most circuits, sync fed to a grid is positive-going, and to a plate, negative-going.

You can tell if there is any sync in the vertical circuits by rolling the picture down with the vertical hold control. If the blanking bar crosses the bottom of the screen with no hesitation, and if you can also roll the picture up smoothly without any stopping or jumping as the blanking bar leaves the screen, there is no vertical sync. Normally, as you roll the picture down, you'll see the blanking bar snap out of sight when it gets 2 or 3 inches from the bottom of the screen. This indicates good sync-lock. If you turn the hold control the other way, the picture should hold to a given point, then break loose and travel upward very rapidly. To distinguish between the two, most technicians call downward movement *rolling* and upward movement *flipping*. In all cases, if you can roll a picture slowly and smoothly upward, there is sync trouble.

Horizontal Sync and AFC Checking

Horizontal sync circuits are just as easy to check as vertical circuits. In many sync circuits, you'll find composite sync at the output of the sync separator: Vertical and horizontal sync pulses are separated by resistor-capacitor networks. The low-frequency vertical sync goes through large capacitors, with good-sized bypass capacitors that shunt the higher-frequency horizontal sync to ground; the horizontal sync goes through very small capacitors, which have a high reactance to the low-frequency vertical pulses.

The typical horizontal afc circuit is a phase comparer. The horizontal sync from the TV signal is compared in phase to a pulse from the horizontal oscillator. This pulse is sometimes taken from the oscillator itself, sometimes from a special winding on the flyback. Both signals are usually shaped into a sawtooth waveform by resistor-capacitor networks. Figure 7-29 shows a typical sawtooth found on an afc diode plate.

Check for the presence of both sawtooth waveforms and compare their amplitudes to the values given on the schematic. As we said, horizontal sync works on phase, so the amplitude isn't too critical. However, it must have a certain minimum value before the afc circuit will work properly.

The easiest way to check an afc circuit is by the process of elimination. Take it out of the circuit by shunting the horizontal oscillator grid, etc. Then see if the oscillator will work alone. If so, it will make a single, floating picture on the screen. Now put the afc back; the picture should lock in and stay in sync. If replacing the afc makes the picture go *out* of sync, or if the picture continues to float (no sync), then the afc circuit must be bad.

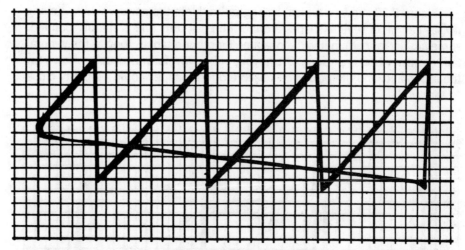

Fig. 7-29. This is a typical sawtooth signal found at the agc diode plate (anode).

If the afc uses a pair of diodes, check them—preferably by substitution, but they can be checked for front-to-back resistance ratio, or shorts, or opens, etc. If the diodes are okay, then check the little sync-coupling capacitors, the 82 to 100-pF micas. These capacitors have been known to leak; if they did, they could upset the dc balance of the afc circuit. Also, check all resistors and capacitors. There are only six or eight of them in an afc circuit and they must all be good if the set is to work properly.

CHECKING KEYED OR GATED STAGES

Only one thing distinguishes a keyed or gated stage from others. A keying pulse is applied to it from the flyback—usually to the plate, but it could be on the screen or control grid, etc. and still do its job. This pulse is the plate voltage in most gated stages, such as 6BU8 agc-sync stages. The pulse is usually positive going and runs from 400 to 700 volts peak.

To check out a keyed stage, first measure all of the dc voltages; make certain they are within 5 percent of the values given on the schematic. Next, check for the presence of the keying pulse on the plate and note its amplitude. Although a given stage might go into conduction at 500 volts, a pulse that should be 700 volts peak and isn't usually indicates trouble elsewhere in the set.

The pulse is shaped into a steep-sided form by a resistor-capacitor network. If the capacitor becomes leaky or the resistor changes value, the pulse shape or amplitude changes and you have trouble.

Figure 7-30 shows a typical keying pulse. Note the narrow width. Keying pulses can be of several shapes around the peak (this one is sharply pointed, some are flat-topped), but variations aren't too important as long as the general tall, narrow shape is present. The keying-pulse waveform is usually shown on the schematic, and the pulse you observe on the scope should be pretty close to that.

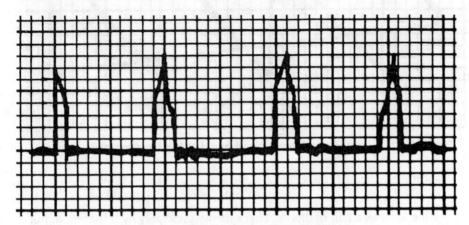

Fig. 7-30. This is a typical keying pulse.

TV SIGNAL TRACING AND GAIN CHECKS WITH AN RF OR AF SIGNAL GENERATOR

You can check out a video-amplifier stage, video-detector stage, or video i-f stage with an rf or af signal generator. The audio output from the rf signal generator can also be used if you don't have a separate af generator. Figure 7-31 illustrates the pattern when a square-wave audio signal (about 600 Hz) is fed into the input of the video amplifier of a black-and-white TV set. Notice the sharp horizontal bars on the screen. (This test can also be used to adjust the vertical linearity—note the compression near the top of the screen.)

The contrast of the bars is used to check out the gain of the video amplifier. The input signal level should be 2 to 3 volts p-p in single-stage video amplifiers like this one. In color TV video-amplifier stages, check the schematic for the average level of video output. If a low-level signal is needed and the attenuator of the signal generator won't go down that far, you can make up a simple resistive voltage divider as described earlier. Don't overload the input, especially in the video stages of transistor TVs.

A scope pattern of the signal in Fig. 7-31 is shown in Fig. 7-32. Notice that there is some distortion of the square wave. If a video amplifier has a very wideband output and if test conditions are just right, you see an almost perfect square wave—up to about 10 kHz, with sharp, clean bars on the CRT. However, the set used for this test produced a very acceptable picture, so don't be overly critical. You can often improve the scope pattern at the output by taking off the picture-tube socket and picking up the signal at the grid or cathode pin of the socket. This substitutes the input capacitance of the scope for that of the picture tube and sharpens the pattern. The pattern of Fig. 7-32 was made with the CRT still hooked up.

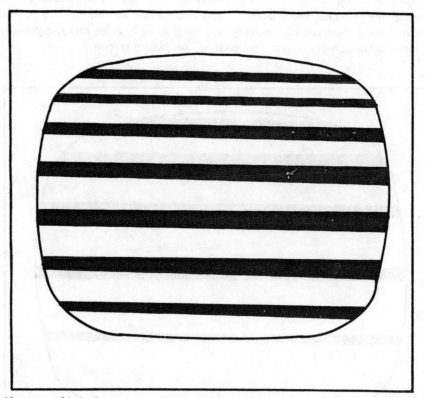

Fig. 7-31. This is the pattern when a 600 Hz square-wave signal is fed to the input of the video amplifier.

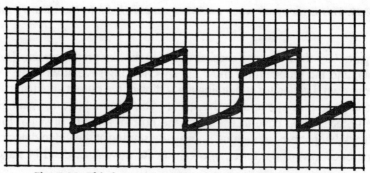

Fig. 7-32. This is a scope pattern of the signal in Fig. 7-31.

A sine-wave signal can also be used. In fact, you can feed an amplitude-modulated rf signal into the input of the set's video i-f stages and see the resultant on the screen, as in Fig. 7-33. (The contrast of the bars also gives you a good idea of the condition of the picture tube.) This is a valuable shortcut test when the big question is whether the trouble is in the video i-f or in the tuner. If an rf signal at the picture i-f frequency (modulated by a 400-Hz sine wave) passes through the i-f stages, video detector, and video amplifier and makes a pattern on the screen, then you can be fairly sure the trouble is in the tuner, especially if the original complaint was "no picture" or "white screen."

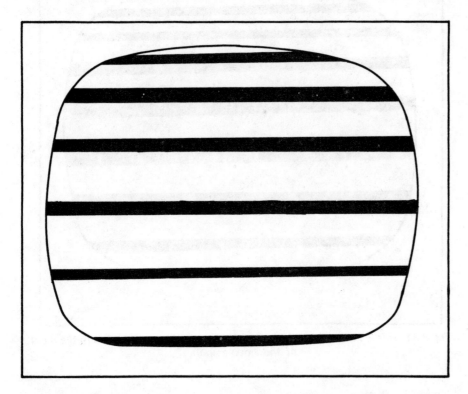

Fig. 7-33. This is the result when an AM rf signal is fed into the video i-f stages.

Figure 7-34 shows what a weaker signal looks like on the screen. The bars are pale, and the retrace lines show up. By checking several working TV sets with your own rf generator and noting the setting of the output attenuators when you get a good black pattern on the screen, you can make rough gain checks on the i-f, etc. If you have to use twice the normal amount of signal to get a good black-bar pattern, something is weak. By starting at the video detector and working back toward the tuner, you can locate a weak or dead i-f stage.

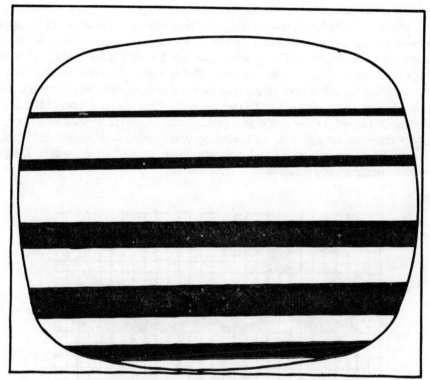

Fig. 7-34. A weaker signal looks something like this.

In fact, you can check video-detector diodes with this test. If an i-f signal fed into the input produces very pale bars, but an audio signal fed into the video-amplifier grid produces good, sharp bars with plenty of contrast, then the video-detector diode could be leaky or open. When making this test, check the setting of the agc control. If the agc is too negative, it will cut the i-f gain and make the picture weak and washed out.

SIGNAL TRACING WITH THE SCOPE AND COLOR-BAR PATTERNS

If you run into trouble in the video i-f or video-output stages, signal-trace through these circuits with the scope to find where the signal stops or loses gain. If we have an easily identified pattern, the job will be easier. Such a pattern can be supplied by the signal from a bar—dot generator set for the color-bar signal or the crosshatch. In circuits where the signal is still rf, such as the video i-f, you need a crystal-detector probe on the scope; after detection, you can use the direct probe, although a low-capacitance probe is handy in many cases.

The main benefit in using this signal is that it can be readily identified.

Figure 7-35 shows a crosshatch signal, taken from a video amplifier through a low-capacitance probe. Figure 7-36 shows a color-bar signal at the same point, taken with a direct probe. The square pulse near the right is the horizontal sync bar of the signal. The set used for these photos was a black-and-white TV, so the color-bar signals are not so plain as they are in color sets. Nevertheless, the characteristic shape of the signals can be seen. Most bar, dot, and crosshatch signals, displayed on the scope at a vertical rate of 30 Hz, resemble a comb. Check them out on a set that is working, so you'll know what they should look like. So far, be mainly concerned with amplitude rather than waveform. Don't be too critical of any distortion, at least not yet.

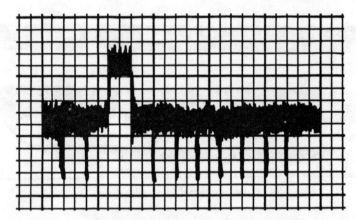

Fig. 7-35. This is a crosshatch signal taken from a video amplifier through a low-capacitance probe.

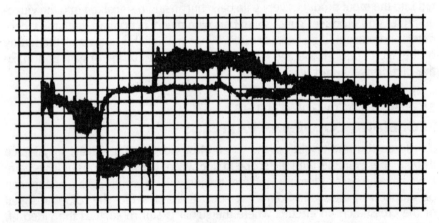

Fig. 7-36. This is a crosshatch signal taken from a video amplifier through a direct probe.

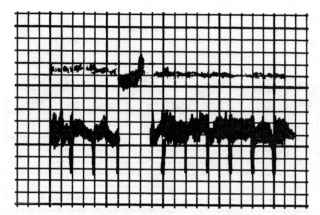

Fig. 7-37. A test pattern is useful for detecting sync clipping.

However, a test pattern can be useful for detecting sync clipping. An example is illustrated in Fig. 7-37.

USING THE SCOPE TO SET THE DUTY CYCLE OF A HORIZONTAL-OUTPUT TRANSISTOR

Correct adjustment of the oscillator-drive waveform in a transistor TV is not as simple as in tube types. The wrong adjustment can cause trouble in a hurry—even faster than in tube sets.

The horizontal-output transistor actually works as a switch; it is driven by a rectangular pulse waveform. In such sets as the RCA KCS-153, you must use the scope to set the ratio of off-time to on-time (duty cycle) in this waveform.

If the on-time is too long, average current will go up. This can cause overheating of the junction and even blow the transistor or kick out the circuit breaker. By using the scope to display the waveform, measure the ratio between on-time and off-time directly.

Here's the complete procedure. It is similar to the procedure used in the old, faithful synchroguide circuit. Two stabilizing coils are used. One controls the oscillator off-time, and the other (which is actually resonated at about 40 kHz) controls the on-time. To make the preliminary setup, ground the collector of the sync-separator transistor and hook a jumper across the sine-wave coil. Now, turn the set on and adjust the hold control for most stationary picture. The picture floats since there's no sync or stabilization, but you can get a single picture by juggling the hold control.

The bottom core of the coil (the 40-kHz section) is adjusted for a pulse-width ratio of 1:2 (1 on, 2 off). The waveform is shown in Fig. 7-38. The wide pulses should be at least twice the width of the narrow ones. If you want the exact recommended figures, the narrow pulses (downward-going) are 18 μs, and the wider off-pulses are 60.5 μsec. As long as the ratio is greater than 1:2, it's fine. You can measure this by setting the pattern width on the scope so that you can measure the two pulses on the calibrated screen.

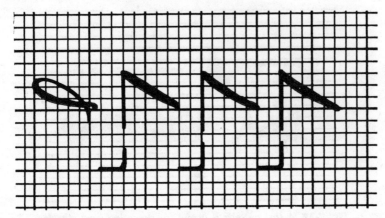

Fig. 7-38. The bottom core of the coil is adjusted for a pulse-width ratio of 1:2.

Next, turn the set off and remove the jumper across the sine-wave coil. (Do not connect or disconnect anything in a transistor circuit with the power on. It's dangerous. Making or breaking a connection could cause transients that would puncture transistors. Turn the set off!) Turn it back on and adjust the top core—the sine-wave coil—for a locked-in picture. Remove the scope and the job is done. Disconnect the jumper that grounds out the sync, too, before you forget it.

HOW *NOT* TO READ VOLTAGES:
THE "DO-NOT-MEASURE" POINTS IN TV

Now that you know how to read dc voltages at typical points of a TV circuit, let's see about those points that must *not* be measured—the plate of the horizontal output tube, the high-voltage rectifier plate, the damper tube, and the vertical output plate. These points should be marked "Do Not Measure" on the schematic.

They say "Do Not Measure" because there is a very high *pulse* or *spike* voltage present at all times, besides the dc plate voltage. During normal operation, these spikes reach 15,000 to 20,000 volts peak. Such spikes of voltage can damage the multiplier resistors in your VOM by causing a flashover (internal arcing) in the precision resistors. These resistors are seldom rated at more than 1 watt—usually less. Therefore, VOMs should never be subjected to this kind of mistreatment. Once a multiplier has flashed over, the meter will be inaccurate until it is recalibrated.

You *can* read the dc voltages on these TV circuit points if you use the right methods. In each case, the voltage is fed through a coil—the horizontal or vertical output transformer, etc. So, turn the set off and read the resistance of the coil primary. Check this against the value given on the schematic. Now, turn the set on, and read the dc voltage at the *bottom* of the coil—the opposite end from the plate of the tube. This is where the supply voltage is fed in; if it's okay there, and if the coil winding shows continuity, there *is* plate voltage on the tube in question.

BLOWN AUDIO OUTPUT TRANSISTORS

Audio output transistors seem particularly susceptible to blowing out. Many technicians have run into problems where they replace a blown audio output transistor, and the new one immediately blows out when power is applied. This clearly indicates that the blown transistor is just a symptom of some other problem.

But how can you locate the problem without being able to take any voltage or current measurements? The solution is to use a Variac. Before applying power, plug the set into the Variac and adjust its output to zero. Now turn on the TV, and monitor the current flowing through the audio output transistor as you slowly increase the voltage from the Variac. Many schematics indicate how much current should be flowing through this transistor.

Generally, TV audio stages operate in the class-B mode. In this type of circuit, the quiescent (no input signal) current drain should be considerably less than the full-load current drain. If you detect any appreciable current drain with a low supply voltage, trouble is clearly indicated.

DEALING WITH DEAD SETS

To the average layman, the ultimate equipment catastrophe is a completely dead set. No raster, no sound, no lighted indicators, no nothing.

However in most cases, a completely dead set is the easiest to service. It is certainly less frustrating to work on than one with an intermittent problem.

When faced with a completely dead set, the first step is to check the painfully obvious. Is the set plugged in? You might be surprised to learn how many service calls are made for unplugged equipment. Even on my own work-bench I have embarrassed myself by being momentarily confused by something I forgot to plug in or a power strip I forgot to turn on. I suspect any honest technician will admit to making such a goof at some point in his career. We're only human, and it's all too easy to make assumptions.

Even if the set is plugged in, that is no guarantee that power is being applied to the set. Is the socket live? Is there a blown power line fuse? Sometimes just a single socket goes dead, so always try plugging something else into the questionable socket before drawing any conclusions, for example a lamp. Many technicians carry electrical socket testers in their tool kits. These devices are handy and quick. Besides giving a clear indication of power, they also indicate whether or not the socket is properly grounded. These simple testers only cost a few dollars.

If power is being applied to the set, check the power cord. There could be a small break in it. Many sets have detachable power cords. Make sure the cord is fitted into the set properly and securely.

Better quality sets normally have internal fuses, or more commonly, circuit breakers. If the fuse is blown, obviously it should be replaced. If the circuit breaker is tripped, reset it. Then let the set run for a while. The problem could have been caused by a brief transient in the power lines, but there could be something wrong within the set's circuitry. Generally speaking, if the problem is still present, the fuse or circuit breaker will soon blow again. In some cases it blows instantly as soon as power is applied.

Servicing this kind of problem can be rather tricky. After all, you can't take any voltage or current measurements. Do *not* try bypassing the fuse or circuit breaker. Remember, it is blowing for a reason. Without the current protection of the fuse or circuit breaker, other possibly expensive components could be damaged or even destroyed. There might be a risk of fire hazard.

Try using your nose and eyes. Does something in the circuit smell or look burned? Especially look for discolored resistors that might have changed values. If anything looks even remotely suspicious, take a passive resistance reading. You might have to lift one end of the questionable component from the circuit to get an accurate or meaningful reading.

Check the wiring closely for any potential short circuits. Once again, the ohmmeter can be useful to eliminate any doubt. Watch out for frayed insulation on any jumper wires or bare leads that might have gotten bent from their original position. Occasionally, a small glob of solder can break free and rattle about, eventually shorting something out.

Initially, it makes sense to confine your suspicions to the power supply itself. To determine if the power supply is in fact the culprit, disconnect it from the set's other stages. Apply power to the supply circuit without any load. If the fuse or circuit breaker blows, the power supply is definitely at fault.

If the power supply seems to work okay without any load, you might want to try it with a dummy load. Sometimes a power supply will work fine unloaded, but the fuse/circuit breaker will blow when the supply attempts to drive a load.

If the power supply seems to work fine when disconnected from the rest of the set, try reconnecting other stages *one at a time*. When the fuse/circuit breaker blows, the troublesome circuit is on line.

When the problem is in the power supply itself (which it will be more often than not), a good test procedure is to unplug the set and discharge all of the large capacitors. Then measure resistance from various points in the power supply circuit to ground. You should find fairly high resistances to ground throughout the circuit. In most cases, the resistance should be at least 50K, if not more. A resistance reading of zero, or close to it, almost surely indicates a problem spot. A moderate reading—say, about 15K to 25K, or so—indicates that you are close to the trouble spot, but haven't quite pin-pointed it yet. Shorted capacitors are common causes for such problems.

In some sets, the power supply will be working fine, but the set is still completely dead, except perhaps for a power indicator light. This can certainly be a frustrating problem, but at least you can take voltage measurements. Dc voltages will generally be the most useful in this circumstance. If you find a voltage that is significantly incorrect or missing altogether, a problem is definitely indicated somewhere in that stage of the circuit. Find and correct the problem(s) as in any other repair job.

Work your way through the set's stages from input to output until you turn up the trouble. If the dc voltage tests are unrevealing, go back to the beginning and try ac signal tracing.

8
Special Tests

This chapter includes a variety of unique tests that require special techniques with common equipment.

MEASURING PEAK VOLTAGES WITHOUT A VOLTMETER

Frequently, you need to know the peak-to-peak drive signal voltage—on horizontal output tubes, vertical output tubes, oscillators, etc. Normally, you read this voltage on a calibrated scope or with a peak-to-peak-reading ac voltmeter. However, if these instruments aren't available, you can always use the low milliampere range or the microampere range of the VOM.

Open the bottom end of the grid resistor and hook the microammeter in series with it. Connect the positive terminal of the meter to ground, and the negative terminal to the resistor. With the circuit in operation, multiply the current you read by the value of the grid resistor. The result, according to Ohm's law ($E = IR$), is the peak value of the grid voltage. Note that we said peak, not peak-to-peak; the grid conducts current only on the positive-going halves of the drive signal. So, to get the peak-to-peak value most often specified in service data, double the value calculated.

In one actual test, a 6DQ6 tube with a 470K grid resistor read a bit over 100 μA. This gives 47 volts *peak*; the *signal*, measured with a calibrated scope, was about 90 volts peak-to-peak. Resistors in such circuits aren't precise; tolerances are 10 to 20 percent. Voltages obtained by this method won't be exact; but they will be accurate enough to give the information needed pertaining to the drive signal voltage. Remember to double the reading for a peak-to-peak value.

USING A PILOT LAMP FOR CURRENT TESTING

When you encounter one of those jobs that runs along fine then suddenly blows the fuse, it can be expensive as well as annoying. Fuses cost money, especially if you blow four or five of them before finding the short. Make up a test adapter with a pilot-light socket wired across a blown fuse, as shown in Fig. 8-1. Use a pilot lamp that has a rated current a little higher than the normal drain of the circuit.

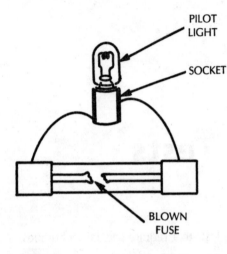

PILOT LIGHT

SOCKET

BLOWN FUSE

Fig. 8-1. A simple pilot lamp can be used for current testing.

For instance, if you're checking the B+ supply of a TV set rated at 200 mA, use a 250-mA pilot lamp—No. 44. There are pilot lamps with current ratings all the way from 150 mA (No. 40, No. 47) to 0.5 amp (No. 41, etc.). There are special lamps rated as low as 60 mA (No. 49) if you happen to need them.

Now, when you turn the set on, the pilot lamp will glow a medium yellow. You can watch this while you tap, move, heat, or test parts in the circuit. If you hit something that is causing the short, you'll see the lamp flare up to a bright blue-white. A dead short will blow the lamp out, of course, but pilot bulbs are usually cheaper than slow-blow fuses!

As a matter of fact, Motorola and other two-way radio manufacturers have used pilot lights as B+ fuses in power supply for several years. If you run into one of these circuits, be sure to use the same type of lamp as a replacement; it must have the correct current rating.

A QUICK TEST FOR AUDIO POWER OUTPUT

There's a quick and dirty test for audio power output if you don't want to bother with calibrated resistors. Get a weatherproof lamp socket and put a couple of heavy terminal lugs on the wires—spade lugs are best. Connect this socket across the high-impedance output taps on the output transformer of the PA system, as illustrated in Fig. 8-2. Screw a standard incandescent lamp into the socket of a wattage to match the power you want to check. For example, on a 50-watt amplifier, use a 50-watt lamp; for a 30-watt amplifier, use a 25-watt lamp (the nearest commercial size).

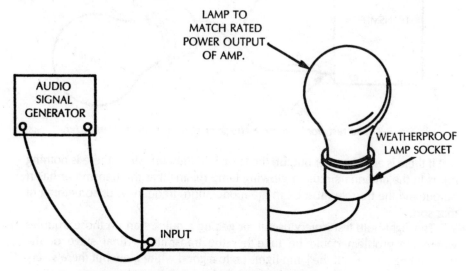

Fig. 8-2. The test hook-up for a check of audio power output.

Now fire up the amplifier, and feed in an audio signal. If you have the rated power output, the lamp will light. If a 50-watt amplifier will light a 50-watt lamp to a good white or about normal brilliance, that's it.

The impedance match is closer than you'd think on the 500-ohm output. A 25-watt lamp, for instance, draws about 0.21 amp; plugging this value into the formula $W = I^2R$ gives a hot resistance of about 570 ohms, which is fine for all practical purposes. If you want, you can parallel two 25-watt lamps across a 250-ohm output and get a fairly close match on a 50-watt amplifier. This method isn't good for high-power transistor amplifiers, because more critical impedance-matching is needed. This would be especially true in the output-transformerless types; for those, use the exactly matched load resistor mentioned in the earlier test.

A QUICK TEST FOR THE RF POWER OUTPUT OF A TRANSMITTER

An incandescent lamp of the proper wattage rating makes a good quick-and-dirty test for rf power output in high-power transmitters (anything from about 10 watts up). Get a weatherproof lamp socket, a rubber-covered type with pigtail leads, and connect it to about 18 inches of coax with a PL-259 plug on the end. If you find a transmitter where there is doubt about the actual power output, screw this plug onto the antenna socket, put a lamp of the appropriate wattage in it, and key the transmitter. This is shown in Fig. 8-3.

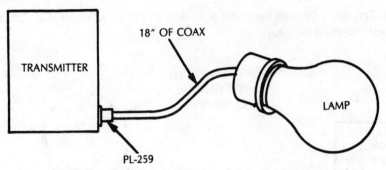

Fig. 8-3. The test hook-up for a check of rf power for a transmitter.

If there is any rf power output, the lamp will light up. Since there is nothing but rf in the antenna circuit, a glowing lamp means that the transmitter has rf output and the trouble must be in the modulation, frequency, or something of that sort.

This light-bulb test is very useful for getting a quick start on those troubles where the problem could be in either the transmitter's final stage or the transmitting antenna. If the lamp lights up to a good white glow but there's very little output from the antenna, then check the antenna and transmission line; that transmitter's okay. Although the mismatch between the transmitter and the lamp is theoretically something awful, you'll be surprised how little you have to change the tuning adjustments between the lamp and the regular antenna.

For CB transmitters? Use a No. 47 pilot light. Even lower power? Try a No. 49.

MEASURING BASE BIAS
VOLTAGE IN HIGH-RESISTANCE CIRCUITS

In a few audio-amplifier applications, you'll find very high-value resistors used in the base voltage-divider circuits. One well-known make, for example, uses a 12 M resistor as part of the network. Even the 11 M resistance of a VTVM will upset this circuit if it is shunted across. Since it takes only a fraction of a volt to cause a big change in transistor currents, measuring the base bias directly is not a good procedure.

Setmakers recommend measuring collector and emitter voltages as given in the service data. If these are correct, then the base bias must be right. In other words, avoid getting into the very high-resistance circuit to measure base voltage. Instead, measure the voltages that this voltage affects.

If you want to check out the base-bias circuit, remove the transistor and measure the large resistors for proper value, especially if the collector and emitter voltages are off-value.

A VOLTAGE DIVIDER FOR
OBTAINING VERY SMALL AUDIO SIGNALS

For audio-amplifier testing, you need a source of low-level signals. Be very careful not to overload the input, in transistor amplifiers especially. Also, it is possible to make a quick check on any audio amplifier by feeding in a given number of millivolts and measuring the audio-output power. If the amplifier comes up to specifications, there's no need to go any further. This is also a fast way to isolate a weak channel on a stereo system.

You can read the output of an amplifier in watts by substituting a load resistor of the proper value for the speaker, applying a steady signal to the amplifier input, and measuring the voltage across the load resistor. The value of power output can be figured out by the Ohm's law equation $W = E^2/R$. In many cases, the right values of voltage for both input and output are given in the service data.

For instance, an amplifier might call for 2.75 volts across the output-load resistor at an input of 300 mV. The problem is to get a true 300 mV since most shop-type audio generators don't have accurately calibrated attenuators. To get the signal needed, use a simple resistive voltage divider, as shown in Fig. 8-4. If the upper resistor has a value of 700 ohms and the bottom resistor is 300 ohms and you apply exactly 1 volt of audio signal, you can take off 300 mV at the tap.

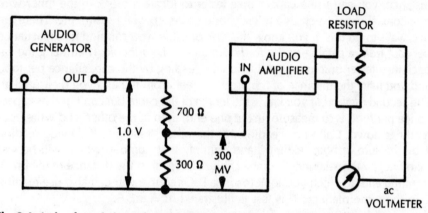

Fig. 8-4. A simple resistive voltage divider can be used to bring the signal down to the desired level.

It might be more convenient to use a 1,000-ohm variable resistor, as in Fig. 8-5. Hooked up as a potentiometer, the slider can be set with an ohmmeter to give any value of signal output needed. Stereo and hi-fi amplifiers designed for low-output magnetic cartridges usually call for about 5 to 10 mV, while "ceramic" inputs call for 700 mV to 1 volt. With the signal-generator output set at a given level, the pot can be used to obtain any fraction of this amount. This method eliminates the need for accurate measurement of very low signals; all you need to know is the total signal across the divider and how the divider is set up. A carbon potentiometer is best. The inductance of a wirewound pot could affect the accuracy of division at high frequencies.

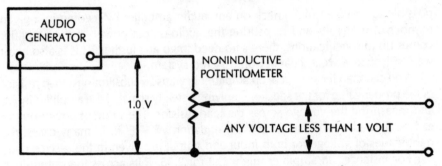

Fig. 8-5. It might be more convenient to use a 1K potentiometer.

FINDING A BREAK IN A COAXIAL CABLE WITH A CAPACITANCE TESTER

You can use almost any piece of test equipment for many applications. For instance, you can use a capacitance tester to locate a break in the inner wire of a coaxial cable, such as a microphone cable, coaxial lead-in, etc. There are two ways to do this. If you know the type of cable, you can find its capacitance per foot from a catalog or the *Radio Amateur's Handbook*. Then hang the capacitance tester onto one end. Divide the reading by the capacitance per foot, and you have the distance of the break, in feet, from the end you measured at. The second method, if you know the length of the cable but can't find its capacitance per foot, is to measure first at one end, then at the other, and write both readings down. Call one "reading 1", the other "reading 2". Divide reading 1 by the sum of both readings, and multiply that number (which will be less than one) by the total length of the cable. The result is the distance of the break from the end at which you took reading 1. Of course you can also use reading 2 in the same manner. This test is illustrated in Fig. 8-6.

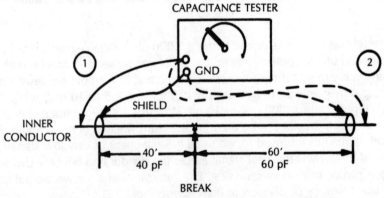

Fig. 8-6. A capacitance tester can be used to find a break in a length of coaxial cable.

Most breaks in microphone cables happen near one end or the other because of the bending and flexing near the plugs. To find out which end is broken and save tearing the plugs apart needlessly, take a capacitance reading from each end. One end will show a large capacitance, the other practically none; the latter is the broken end.

A capacitance test can also be used to find a broken wire inside the insulation of a two-conductor cable. Use the good wire as the "shield" and measure the capacitance of the broken wire to this. Otherwise, the method is the same.

MEASURING TRUE RMS VOLTAGES

Measuring dc voltages is generally straightforward. Five volts is five volts. The only possible source of confusion is the reference point for the voltage measurement. Normally, the circuit's ground potential is used as the zero reference point, so this is rarely a problem. The few exceptions are usually clearly indicated.

However, things get more complicated when you start dealing with ac voltages. An ac voltage, by definition, is constantly changing its value from instant to instant.

There are several different ways to measure ac voltages, which are useful in certain circumstances. For instance, in some cases you are concerned with absolute maximum values. Here, you want to know the peak voltage reached during the ac waveform's cycle. For convenience, assume all waveforms are centered around zero volts dc.

An ac voltage might be in the form of a sine wave with a peak voltage of 10 volts. (It would also have a peak negative voltage of −10 volts.) To say you have 10 volts ac would be misleading, because the peak value is reached for only a tiny fraction of each cycle. For most of the cycle, the instantaneous voltage is considerably less than the peak value.

Because the voltage varies between +10 volts and −10 volts, it covers a 20-volt range. In some instances, it is useful to measure the signal as 20 volts peak-to-peak. But once again, saying there's 20 volts ac would be misleading.

The logical approach would be to take an average of the instantaneous voltages throughout a cycle. However, you can't use the entire cycle, because for a symmetrical waveform the positive half-cycle and the negative half-cycle will always cancel each other out, leaving an average value of zero volts.

For average ac voltages, only half of each cycle is considered. It has been mathematically proven that for a sine wave, the average voltage is always equal to 0.636 times the peak value. In our example of a sine wave with a peak voltage of 10 volts, the average value would be 6.36 volts.

This is not unreasonable, and average values will allow us to meaningfully compare various ac voltages. Unfortunately, the formulas of Ohm's law do not hold true for average voltages. This is a major loss to the electronics technician, because so much of circuit design and analysis is based on the relationships described by Ohm's law.

What you need in order to retain Ohm's law for ac voltages is a way to express ac voltage in terms that can be directly compared to an equivalent dc voltage. Such an equivalent value can be found by taking the *root mean square* (rms) of the waveform. For a sine wave, the rms value is always equal to 0.707 times the peak value. In the example, the 10 volts peak ac sine wave would be measured as 7.07 volts rms. This signal would heat up a given resistor exactly the same amount as 7.07 volts dc through the same resistor. The voltage/current/resistance relationships defined by Ohm's law work out the same for rms values as for dc values. As a rule, ac voltage and current relationships are usually measured as rms values.

I have gone into some depth here because even experienced technicians occasionally get confused. For your convenience, here is a handy comparison of the various ac measurements.

RMS	= 0.707 × PEAK
RMS	= 1.11 × AVERAGE
AVERAGE	= 0.9 × RMS
AVERAGE	= 6.36 × PEAK
PEAK	= 1.41 × RMS
PEAK	= 1.57 × AVERAGE
PEAK-TO-PEAK	= 2 × PEAK

Table 8-1. Voltage Relationships Between Peak, Peak-to-Peak, and Rms.

First, you have to convert an ac value from one form to another. It is vitally important to remember that these equations hold true *only* for sine waves. They cannot be used for other waveforms.

Most ac voltmeters are calibrated to measure ac voltages for sine waves. These meters will not be accurate for ac signals with any other waveshape. You are usually only concerned with making simple comparisons, so an ac voltmeter should suffice even for non-sine-wave signals. However, if you need a true rms value for a non-sine-wave signal, a standard ac voltmeter is virtually useless.

Moreover, standard ac voltmeters are frequency dependent. They are tuned for accurate measurement at the standard line frequency (60 Hz).

These meters do the job fine for at least 75 percent of all service jobs. But there are occasions when they fall short. Until fairly recently, the technician didn't have much choice in the matter. If a standard ac voltmeter couldn't do the job, then you settled for compromised measurements, estimating, and making educated guesses.

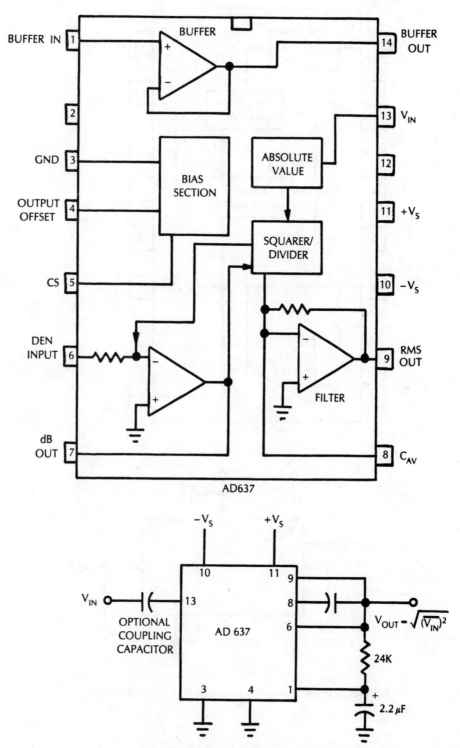

Fig. 8-7. The AD637 is a rms-to-dc converter IC.

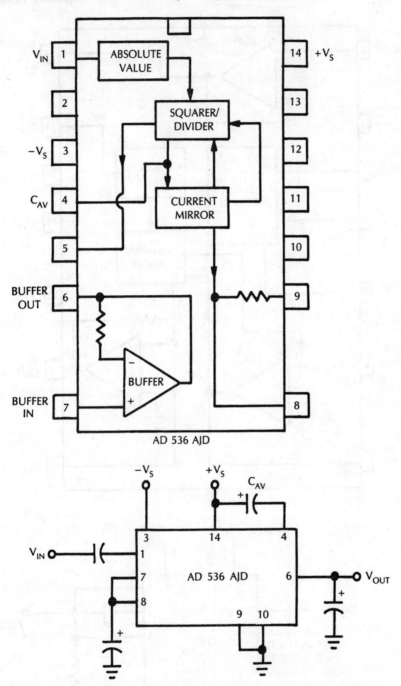

Fig. 8-8. Another rms-to-dc converter IC is the AD536AJD.

Nowadays, there are computer-driven meters that can perform the necessary mathematical calculations. Even better, the last few years have seen the development of specialized ICs that perform rms-to-dc conversion. An ac signal with almost any waveshape is fed into the input, and a proportionate dc voltage appears at the output. This dc voltage is the same as the true rms value of the input signal. These chips are starting to be used in multimeters, especially those of the digital variety.

Two of the first rms-to-dc converter chips are the AD637, shown in Fig. 8-7, and the AD536AJD, which is illustrated in Fig. 8-8.

TESTING TRANSISTORS

The vast majority of transistors are of the standard bipolar variety. *Bipolar* means that the semiconductor contains two pn junctions. The bipolar transistor is essentially a semiconductor "sandwich", as illustrated in Fig. 8-9.

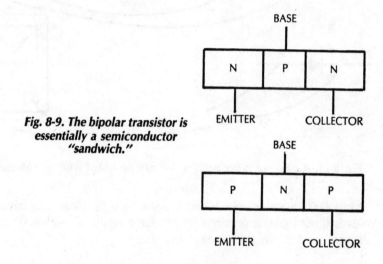

Fig. 8-9. The bipolar transistor is essentially a semiconductor "sandwich."

Functionally, the bipolar transistor can be considered a pair of back-to-back diodes interconnected as shown in Fig. 8-10. This is just an illustrative simplification. Normally, you cannot replace a transistor with two discrete diodes.

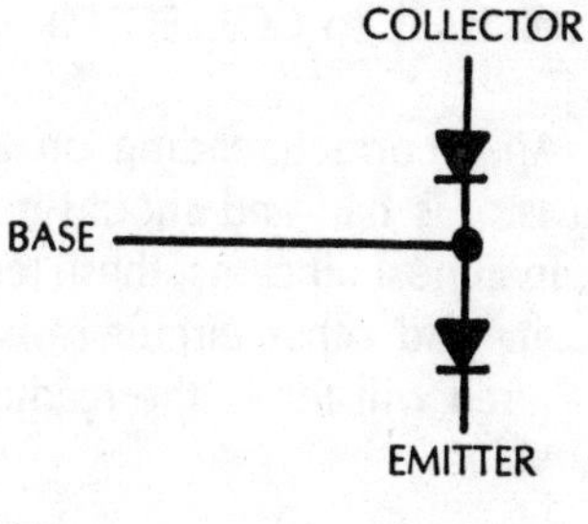

Fig. 8-10. Functionally, the bipolar transistor is like a pair of back-to-back diodes.

A diode can easily be tested with an ohmmeter. Simply measure the resistance from one lead to the other, then reverse the leads and measure the resistance again. This process is illustrated in Fig. 8-11. You should measure a very high resistance in one direction and a low to moderate resistance in the opposite direction. If the resistance is high no matter which way you position the meter leads, the diode is open. If you measure a very low resistance in both directions, the diode is shorted.

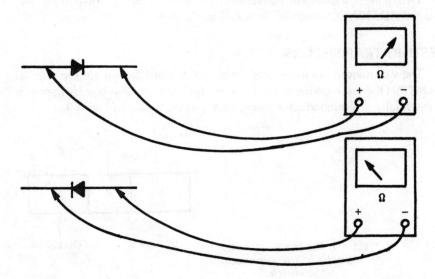

Fig. 8-11. A diode or other pn junction can be tested with an ohmmeter.

A bipolar transistor can be tested in a similar way. Measuring between any two leads should give approximately the same results as with a single diode. In all, you need to make six measurements:

- EMITTER to BASE
- BASE to EMITTER
- EMITTER to COLLECTOR
- COLLECTOR to EMITTER
- COLLECTOR to BASE
- BASE to COLLECTOR

An incorrect reading on any of these measurements indicates that the transistor is bad and should be replaced.

In almost all cases, these tests should be made with the transistor out of the circuit. And other circuit resistances in parallel with the pn junction being measured will throw the reading off, especially if the parallel resistance has a low value.

The ohmmeter test works, but it is a little tedious, swapping the test leads around six times. A dedicated transistor tester is a lot more convenient. Some very fancy transistor testers are available that measure transistor parameters such as alpha and beta. These devices can be useful in certain cases, but for most repair work, you only need a simple "go/no-go" type of test. Many simple circuits have been designed for this purpose. Perhaps one of the simplest is the one shown in Fig. 8-12. You could easily build such a circuit in less than a half hour. The odds are good that you may already have all the necessary components handy. The circuit is powered from an ordinary 9-volt transistor radio battery.

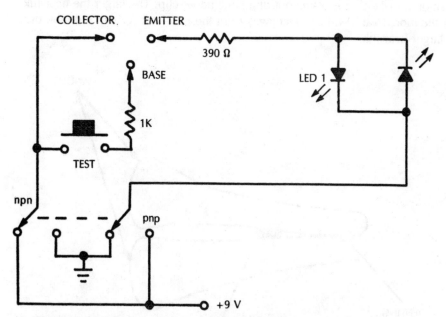

Fig. 8-12. This is a circuit for performing simple "no/no-go" transistor tests.

The transistor to be tested is inserted in the socket. Alternatively, if you prefer, you can replace the socket with test leads that have alligator clips on the ends. The clips are attached to the appropriate leads of the transistor. If the transistor to be tested is npn, the dpdt switch is set to position A; this switch is set to position B for pnp transistors.

The momentary-contact push button is then closed to test the transistor. If the transistor is good, the appropriate LED should light. If the LED does not light, assume the transistor is bad and replace it.

This handy little circuit can also be used to identify a transistor's type. Simply test the transistor with the dpdt switch in both positions. In one position an LED should light, while in the other position both LEDs should remain dark. You can identify the type of transistor by which LED lights. If neither LED lights at either position of the dpdt switch, the transistor is probably bad.

Both of these test procedures are for bipolar transistors only. If you try them on other transistor types, such as FETs, MOSFETs, or UJTs, you will not get correct readings.

These tests work for out-of-circuit transistors, but you will not always get meaningful results if you attempt the test on a transistor wired into a circuit. This is a fairly serious limitation for servicing work. Remember that all semiconductors are heat sensitive. If you're not careful, you can destroy a transistor when desoldering or re-soldering. Always use a heatsink when soldering or desoldering around semiconductor components. The spring-loaded clip-on types, as shown in Fig. 8-13, are usually the most convenient to use. If you don't have one handy, fashion a make-shift heatsink out of a large paper-clip. The larger the heatsink is, the more heat it will conduct away from the transistor. When in doubt, use a larger heatsink.

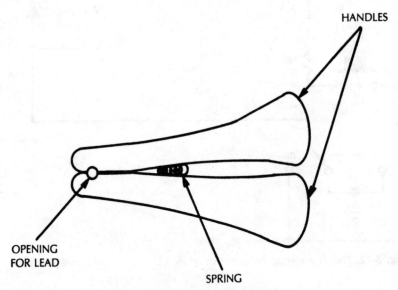

Fig. 8-13. Simple clip-on heatsink.

Obviously it is impractical to desolder every transistor in a piece of equipment for out-of-circuit testing. Such a procedure should be limited to components where there is good reason to assume they're bad.

It usually isn't too difficult to isolate potentially troublesome transistors by making current and voltage tests. Standard values are usually printed in the schematic. Don't worry too much about minor errors. Look for voltage and current values that are significantly off. If a signal being fed into a transistor is correct, and the voltage or current coming out isn't, then it is reasonable to suspect that transistor.

Don't be too quick to jump to conclusions, however. Often a perfectly good transistor can look bad because of a problem in one or more of its associated components (resistors, capacitors, etc.). Even if you remove the transistor and find it tests bad, check the associated components anyway. Almost every electronics technician has had the frustrating experience of replacing a bad transistor only to have the new component soon blow out too. Something else is wrong, causing the transistor to go bad.

Actually, such multiple problems are more common than you might suspect. Semiconductors are pretty reliable. Unlike tubes, transistors seldom go bad by themselves (although it has been known to happen occasionally). Usually something external has caused the transistor to go bad. In some cases, the transistor has been damaged by a transient that somehow managed to get into the circuit. In such an instance, there might be nothing else wrong with the circuit. When the transistor is replaced, the unit will work just fine. More frequently, however, the transistor's problem will have been caused by a resistor that has changed value, an open or leaky capacitor, or something similar.

Problems caused by other components might not always be obvious. Often it takes them a while to damage the new transistor. Run the equipment at least a half hour after making the repair. This reduces the number of call-backs and irate customers you might have to face.

There is one final point about transistors you should be aware of. Unlike tubes, transistors are usually either definitely good or definitely bad. They are rarely weak. Once again, though, there are exceptions, so never be too hasty to jump to conclusions. On those rare occasions when you suspect a leaky transistor, you need a high-quality transistor tester to check it. A VOM/VTVM or the "go/no-go" tester like the one shown back in Fig. 8-12 won't be enough. A silicon transistor can have leakage as small as 10 to 15 microamperes. This small amount of current error is very difficult to measure by standard methods but can be sufficient to throw off proper circuit operation and cause all sorts of strange symptoms.

You should suspect a leaky transistor when everything in the circuit (resistors, capacitors, supply voltages) seems okay, but the circuit simply doesn't work the way it ought to. If all passive components are good, then an active device (such as a transistor) must logically be the culprit.

To perform a leakage test with a transistor, you almost always need to remove the component from the circuit. If you don't have a suitable transistor tester for the leakage test but do have a duplicate of the questionable transistor, try substituting the duplicate in the circuit. If the circuit now works properly, it is safe to assume that the original transistor was leaky and should be discarded. On the other hand, if the symptoms persist, the odds are very strong that the problem lies elsewhere. It is extremely unlikely you'll encounter two leaky transistors (the original and the duplicate) in a row.

TESTING ICS

Integrated circuits, or ICs, can be tricky for the troubleshooter. Essentially, you have to treat an IC as a black box. All that matters are the external functions. The internal circuitry is irrelevant since there is no way to get at it.

Test the supply voltages and the inputs to the IC. If these are correct but the outputs are wrong, it's a good bet the IC is bad and should be replaced. However, sometimes you can be fooled by loading. Occasionally, a problem in a later stage can throw off the reading at the ICs output pin(s). If there is any possibility of this, break the connection between the output pin in question and the later stages of the circuit. If the IC is okay, you should now get a correct reading at the output pin.

Similarly, a bad IC can sometimes affect the value at one or more of its inputs. Again, the way to find out for sure is to break the connection to the appropriate pin and measure the signal that would be presented to the ICs input if the connection was not broken. If this signal is now correct, the IC is bad.

With IC-based equipment, a schematic is even more important than for other types of servicing jobs. The more you know about how the IC is supposed to work, the better chance you have for properly identifying any problems. Your work area should include as many IC data books as possible. Virtually all IC manufacturers supply data sheets, and many offer extensive data books on their products. These are invaluable.

If there is a particular type of IC you deal with frequently, it might be worth your while to rig up a simple dedicated tester for that particular IC. The easiest way to do this is to construct a simple circuit around the IC, using a socket. Now just plug the questionable IC into the socket and apply power. If the circuit works, the IC is good. Like transistors, ICs are usually either good or bad. It is very unusual to find a weak IC, although it could conceivably happen. However, don't consider such a possibility unless everything else has been ruled out. The best approach then is to simply replace the questionable IC with a duplicate unit and hope for the best.

The test circuit should be as simple as possible. For example, I work a lot with op amps, like the 741 and its descendants (which all have the same pin designations). For a tester, I put together a basic oscillator circuit with a small speaker. When I plug a good op amp IC into the socket and turn it on, a tone is produced by the speaker. This is a simple, clear-cut indication that things are working properly.

While on the subject of ICs, we should consider the question of sockets. Some experts swear by them, and others swear at them. IC sockets certainly make life easier on the service technician. ICs typically have fourteen or sixteen pins, and some have even more. Soldering (or resoldering) all of those tiny, closely spaced pins without creating solder bridges or overheating the delicate semiconductor crystal within the IC takes a sure, steady hand and fairly precise timing. It is also a tedious job, especially desoldering.

If a socket can be used, an IC can be popped out and replaced in less than a minute, without even plugging in the soldering iron. The use of a special IC extraction tool is highly recommended to avoid bending or damaging the pins. Some technicians routinely add a socket whenever they replace an IC. Then if a similar repair is ever needed, the job will be considerably easier.

Unfortunately, IC sockets are not an unmixed blessing. (And what is?) The electrical contact between the pins and the traces of the pc board is not quite as good as with soldered components. In the vast majority of circuits, this probably won't make much difference. However, in some applications (especially those involving high frequencies), the use of a socket can lead to erratic operation. In portable equipment that can get knocked around, sockets are often undesirable. An IC can work its way out of its socket. It could get damaged. It could even short out some other part of the circuit, blowing out additional components. Finally, in many pieces of equipment, there just isn't enough room for a socket.

Use your own judgement on whether or not to add IC sockets. Some equipment already has sockets from the manufacturer. In this case, just be grateful for the relative ease of repair. There's no point in removing a socket installed by the manufacturer. (However, a socket added by another technician could conceivably be the source of problems in high-frequency circuits.)

Great care should be exercised when replacing an IC, whether with a socket or not. Make sure the IC is correctly oriented. Applying power to a backwards IC will almost surely result in disaster. Most ICs have a small notch indicating their "front" side, or they have a small dot over pin 1. Before removing an old IC, always make a note of its orientation.

When soldering an IC, be very, very careful. Don't apply too much heat. The pins are very closely spaced, so it is very easy to create an accidental solder bridge between two pins or adjacent pc traces. Never use too much solder. Always heat-sink any IC before soldering it.

Make sure all of the pins go into the holes (of the pc board or the socket). One might get bent up under the body of the IC. If this is not caught and corrected, there is a good chance the circuit will not operate correctly.

It pays to be extra careful when working with ICs. For one thing, many of them are very expensive. This is especially true of recently developed or highly specialized devices. Mass manufacture tends to bring down the cost of general purpose devices. But it's still smart to use extra caution even when replacing a "garden-variety" IC that only costs a quarter or fifty cents. Let's face it, replacing an integrated circuit is a royal pain in the lower anatomy. There's no reason to be careless to only end up having to do it over.

9

Signal Tracing and Alignment Tests

Let's begin by discussing alignment methods for tube radios. The first output meter used for aligning tube radios was an ac voltmeter across the speaker-voice coil. A 400-Hz AM signal was used. This was nice, but noisy. Of course, you can always unhook the speaker and substitute an equivalent resistor. An easier way is to use a VTVM on the avc line. The avc voltage developed across the diode detector is always directly proportionate to the amount of signal. This is a negative-going voltage, and you'll find values from 1 volt to 15 to 20 volts.

Figure 9-1 shows a typical avc circuit as used in tube radios. This one is rather elaborate. You'll find fewer filter resistors and bypass capacitors in the smaller sets, but they all work the same way. The avc voltage appears at the top of the volume control, which is the diode load resistor. In most cases, you can pick up the avc at the mixer-grid section of the variable capacitor, so that you won't have to take the set out of the case. It can be picked up at any point along the avc circuit. Since the avc circuit has a very high impedance, you'll have to use a VTVM to get a readable deflection.

For best results, keep the input signal down to the point where it causes the smallest readable voltage on the avc; 1 volt is a good average value. By doing this, you avoid the danger of overloading rf stages and flattening the response peaks.

ALIGNING FM I-F STAGES WITH
A DC MICROAMMETER IN A GRID RETURN

In aligning AM i-f stages, use avc voltage as an indicator by tuning for maximum negative voltage. In FM i-f stages you run into problems because many of them do not use or need, conventional avc. In fact, signal overload is no problem in FM; the more signal, the better it works. A high-level input signal causes clipping, but this is fine; it clips off noise peaks and makes the system quieter. Limiter stages run saturated for best results.

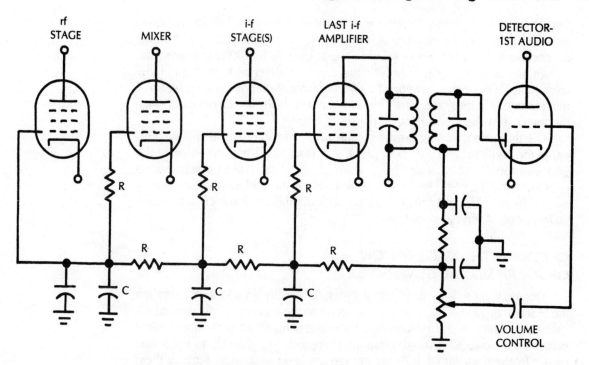

Fig. 9-1. Some older tube radios used an avc circuit like this.

There is one quantity, however, that is usable for alignment indications; this is the grid current drawn by the limiter and even the i-f stages. The harder these stages are driven, the more grid current flows, especially in limiters. The value of this grid current is directly proportional to the amplitude of the input signal, which gives us the kind of alignment indicator we want.

Do not disturb the tuned circuits, but you need to get to this current at some point where test instruments will have no detuning effects. Open the bottom end of the grid resistor, and insert a very sensitive ammeter. A 0 to 50 μA meter will usually be about right. Incidentally, the meter is a common type, used in 20,000 ohms-per-volt VOMs.

With the meter in the grid-return circuit of the last limiter, feed an i-f signal in at the mixer grid. This is an AM signal, remember, and can be either modulated or unmodulated; it makes no difference. Now increase the rf output until you get a readable meter deflection, usually 10 to 20 μA. Now tune all i-f transformers for maximum meter readings.

If you're thinking about the bandpass of the i-f transformers, this is usually built in: Almost all FM i-f transformers have a response curve that is just slightly peaked in the center; it is this peak you tune for. The design of the transformers then gives the proper bandpass. If special i-f transformers are used—for example, triple-tuned types with tertiary windings, etc.—check the alignment instructions and they'll tell you what procedure should be followed. You can make a rough check of bandpass by moving the signal generator dial back and forth either way from the center frequency; this will show you about where the 3-dB-down points are. If you can get a bandpass of at least 40 kHz either side of center, it's okay for cheap sets. Hi-fi FM tuners should have at least 75 to 100 kHz either side.

There's another application for the test just described: measuring the amplitude of the voltage developed by an oscillator. Again, we place a sensitive microammeter in the grid return. This does not detune the oscillator circuit since the meter is in the ground end of the grid circuit.

Use the current reading to estimate the amplitude of the signal voltage present at the oscillator grid. For example, if you read 32 μA and the value of the grid resistor is 470K, that's (32×10^{-6}) (470×10^3), which equals approximately 15 volts. This is the peak value of the grid alternations during each half-cycle. This test can be used to check oscillators for sufficient injection voltage, output, crystal activity, etc.

CHECKING THE CALIBRATION
OF AN RF SIGNAL GENERATOR

Do you want to set your rf signal generator exactly on 455 kHz to realign the i-f stages of a radio? Do you have doubts as to the accuracy of the generator's calibration? Then use the most convenient source of highly accurate test signals—broadcast stations. All AM radio stations are required by the FCC to keep their carrier frequencies within ± 20 Hz of their assigned frequency. Most of them hold to within ± 5 Hz.

To set your generator precisely on 455 kHz, get any radio that will pick up several stations. Its dial calibration doesn't matter; we're only going to use the receiver as an indicator. Listen to a station long enough to determine the call letters and then look up the carrier frequency of the station. Choose a station as close to 910 kHz as you can find. Tune the signal generator to 455 kHz and then zero-beat this with the station carrier, using an unmodulated rf output. Your generator's second harmonic is beating with the station's fundamental.

If you can't find a station exactly on 910 kHz, locate one on each side as close as possible. Check each, note the error in the signal-generator dial. You can use this error to get the dial set on-frequency. For instance, if each station shows that the signal generator is one dial-marking low, then include this same amount of error when you set the generator for 455 kHz.

If you can't find stations close enough to the right frequency, try the third harmonic. At a 455-kHz fundamental, this is 1,365 kHz, so use a broadcast station at 1,360 kHz (all radio stations are on even number 10 kHz apart).

For high-frequency checking, use standard-frequency stations WWV or (in Hawaii) WWVH. You need a communications receiver that covers to 30 MHz, but the actual dial calibration of the radio is not important. These stations broadcast accurate test signals at 2.5, 5, 10, 15, 20, and 25 MHz. You can identify them easily by the 440-Hz beep tone, broken up by ticks at one-second intervals. Incidentally, these two stations also give standard time signals that are used the world over, in case you want to check your watch.

SETTING AN RF SIGNAL
GENERATOR ON A CRYSTAL FREQUENCY

Every now and then we need to set a signal generator to an exact frequency, either for alignment or calibration purposes, or to check the calibration of the signal generator. If we have a crystal that operates at or near the frequency we need, setting the generator is easy. Some signal generators have provisions for plugging in crystals, but a crystal can be used even with a generator that does not have these provisions.

Connect the crystal between the rf output of the signal generator and the vertical input of a scope using a direct probe, as illustrated in Fig. 9-2. Set the rf output to maximum and turn the scope's vertical gain full up. Connect the ground leads of both instruments together as shown. Now tune the signal generator *very slowly* back and forth over the frequency of the crystal. When you hit the exact frequency, you'll see the scope pattern increase in height.

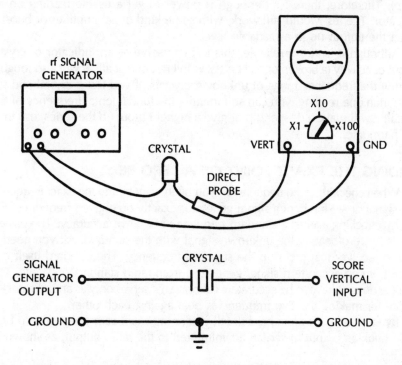

Fig. 9-2. This is a method for setting an rf signal generator on a crystal frequency.

The crystal here is acting as a very sharp resonant filter. When you hit the right frequency, the rf voltage developed across it rises sharply; in typical tests, it may go from .05 to 0.5 volt p-p. The scope need not be a wideband type; all we're looking for here is an increase in pattern *height*. You'll have to tune the signal generator *very* slowly, because the point of resonance is very sharp and you might pass it.

You can use an ac VTVM instead of the scope, or the DC-VOLTS range of a VTVM with a diode in series with the probe. The polarity of the voltage you read depends on how the diode is connected, but this isn't important; all you want is the peak.

You can also hook the crystal up in shunt: Connect the rf output lead and the scope input lead together, connect the ground leads together, and hook the crystal across them. Now you'll see a pattern on the scope (or a voltage on the meter) at all times. When you hit the crystal frequency, you'll see a very sharp dip in the pattern height or in the voltage reading.

When adjusted to the peak (crystal in series) or the dip (crystal in shunt), the signal generator is tuned to the frequency of the crystal and can be used as a standard for aligning rf or i-f stages, receivers, or whatever is necessary. This test works pretty well with crystals up to 4.5-5 MHz, but above this frequency you usually run into low output from signal generators (the average signal generator uses harmonics in the vhf range, thus reducing the output by half or more). Therefore, there isn't enough rf power to get a usable reading on an indicator. The test would still work, with some kind of an amplifier or booster to get the output up to a readable level.

Although it isn't too reliable, this test can serve as an indicator of crystal output or activity (a broken or bad crystal won't respond at all). It will also roughly identify the frequency range of unknown crystals. If you test a crystal and get more than one reading, you can still identify the fundamental frequency of the crystal; it will make a deeper dip or give a higher output (in the series test) than any harmonic.

FINDING THE EXACT POINT OF A ZERO BEAT

When checking radio frequencies, you often want to find the exact frequency of a signal or set the bench signal generator exactly on a given frequency. Do this by checking against a standard frequency of known accuracy. The easiest way is by zero-beating the unknown signal with the standard. All you need is a radio receiver that picks up the standard frequency. The standard itself can come from an accurate rf signal generator, from radio station WWV, etc. The receiver doesn't have to be accurately calibrated; it serves only as an indicator—a device for making the two frequencies beat against each other.

By ear alone, it's often hard to tell where the exact zero point of zero beat is. So hook an output meter (an ac voltmeter) to the radio output, as shown in Fig. 9-3.

Now the signal generator to be checked is coupled to the antenna of the radio, along with the standard. Sometimes just clipping its output lead to the insulation of the antenna lead is enough. Tune this signal to the test frequency, listening for a zero beat in the receiver output as you tune. As the unknown signal approaches the frequency of the standard, you will hear a high-pitched tone that gradually goes lower as you get closer until you can't hear it at all. It then starts to go higher again as you tune past the standard.

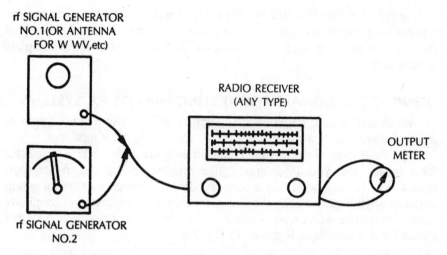

Fig. 9-3. To find the exact point of a zero beat, an ac voltmeter is connected to the radio's output.

Since the beat frequency goes down to zero, below the audible range, the exact zero point is hard to detect by ear. If you want to be very precise, tune for the lowest audible beat frequency, and then start watching the meter. When you reach a very low frequency, you'll see the meter needle start to wiggle as it tries to follow the beat note. At the same time, it will swing slower and slower. Tune for where the needle moves slowest.

TEST RECORDS: GOOD
SUBSTITUTES FOR THE AUDIO-SIGNAL GENERATOR

The average shop needs a high-quality audio-frequency signal generator, but seldom has one. It has to make-do with the 400-Hz audio output of an rf signal generator, although this will do the job for signal tracing and such work.

However, you can get any kind of audio test signal you need at a very low cost, compared to the $300 or more for a high-quality af generator. The source is a test record, and a great many different types are available—stereo, mono, or a combination of the two. With a suitable test record and an inexpensive record player, you're ready for almost any kind of audio work.

A typical test record has single-frequency bands for checking distortion or stylus wear, and a frequency-run series from 30 Hz up to 20 kHz. For stereo, it has left signal, right signal, and both for speaker phasing. Most test records also have many other frequencies. The single-channel stereo signals are very handy for checking separation, etc., and even for identifying channels.

In most cases, you won't even need an audio amplifier; the modern crystal cartridge has an output of up to 3 to 4 volts, and this can be fed directly into many audio circuits. If you want to, you can pick up a small, used amplifier, to supply output signals up to 30 to 40 volts for checking speakers and signal tracing in high-power af stages, etc. This amplifier needn't be hi-fi; you can use it to get the amplifier being tested into working condition, and then feed a signal directly from the cartridge into the input for distortion checks of the complete system with a scope.

USING RADIO SIGNALS FOR TESTING HI-FI OR PA SYSTEMS

When testing hi-fi, stereo, or PA-system amplifiers, use a radio station as a signal source to let the amplifier "cook" for a while after repairing.

The input stage of the average high-gain audio amplifier acts as a detector for rf signals. Hook your TV antenna right across the input, as shown (you must have a closed circuit here). The lead-in and dipole make up a sort of long-loop antenna. Anything like a conical antenna, for example, which has no continuity across the lead-in, won't work; you'd get a tremendous buzz or hum. A simple circuit for this technique is shown in Fig. 9-4.

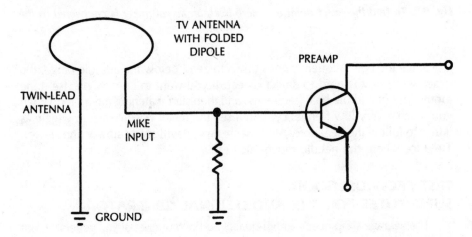

Fig. 9-4. This circuit permits the use of radio signals for testing hi-fi or PA systems.

This technique works best if there's a local radio station near enough to put a good signal into your shop area. However, if the signal is too weak, you can pick up music from a surprising distance by adding a detector diode in series with one side of the lead-in, or in shunt across the input.

If you have several strong locals, this system won't give a single clear sound. An alternative is to pick up the audio from the volume control of a small radio (either tube or transistor). You can set the gain wherever you want it and listen to the music as you cook the amplifier.

USING A COMMUNICATIONS
RECEIVER AS AN RF SIGNAL LOCATOR OR TRACER

Every now and then you need an instrument that can find or trace an rf signal through a circuit, identify an unknown rf frequency, or signal-trace through a circuit to find where the trouble is located. There are special signal-tracing test instruments, but a standard communications receiver (of the kind we used to call a short-wave set) will do the job nicely.

Such receivers have antenna inputs that can be used on dipole antennas or converted to match 75-ohm coaxial cable; a shorting link on the antenna terminal board is used to make the conversion, as shown in Fig. 9-5. Get a piece of RG59/U coaxial cable 3 to 4 feet long and prepare one end to hook up to the terminal board. Cut off about 1 inch of the shield braid on the other end, exposing the insulated inner conductor. Don't take the insulation off; in fact, pull it out over the tip of the inner conductor so the wire can't make contact with anything.

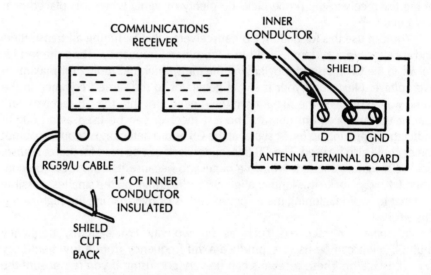

Fig. 9-5. This is a home-made probe for an rf signal locator or tracer.

Now, you can find and identify any rf signal within the tuning range of the receiver. For example, if you're checking a high-frequency oscillator circuit and want to know whether it's working on the right frequency—or working at all, for that matter—simply put the probe end of the cable close to the circuit and tune for the oscillator signal on the receiver dial. The end of the probe cable can be bent into a little hook that can be hooked over wires, etc., near the circuit under test.

An unmodulated rf signal sounds like a little thump in the speaker when tuned across. Tuned directly on it, you hear a rushing sound. A better way to identify it is to turn on the receiver's beat-frequency oscillator (bfo). Now if you cross an rf signal, you'll hear the characteristic beat note or squeal.

For an indication of the strength of the signal, hook a dc VTVM to the test receiver's avc circuit. For convenience, run a lead from the avc to a jack on the receiver panel or back apron, etc., and connect the VTVM there. If the receiver doesn't have avc, use the dc voltage developed across the audio detector; this is usually where the avc voltage comes from anyway. The voltage will go more negative as the input signal strength increases.

A receiver with a probe and an avc meter attached makes a handy little alignment indicator. For example, if tuning up an rf amplifier stage, feed in a signal at the desired frequency, tune the rf stage to it, and then pick up the rf signal output at the mixer grid with the probe. Now, you can adjust the input signal to a very low level to really sharpen the tuning of the rf stages.

You can also follow an rf signal through these stages to determine if it is getting through, if there is a loss or gain in each stage, and to make many tests that are impossible with other test equipment. The probe will not seriously detune any stage, because it should never make actual contact. Because of the sensitivity of the test receiver, the probe picks up plenty of signal when it is placed near any circuit.

You can use this for netting CB transceivers—that is, tuning all transmitters and receivers in a system exactly to each others' frequency. Tune the test receiver to the CB transmitter by keying the transmitter and tuning for maximum avc voltage. Next, tune your rf signal generator to the same frequency in the same way. Now feed the signal from the rf generator into the CB receiver, and tune it up for maximum output. The test receiver can be used as an output indicator by picking up the i-f signal at the CB audio detector. A scope or output meter could also be used. The CB transmitter itself can be used for this of course, but the rf signal generator will do a better job because its output can be controlled. Always make this kind of alignment adjustment on the smallest possible rf signal to avoid flattening the response curves or overloading the high-gain rf circuits.

In some vhf receivers such as in two-way FM systems, frequency multiplication can be used to produce a vhf frequency from a low-frequency crystal oscillator. These receivers can be very confusing if you're not sure the multiplier coils are tuned to the correct harmonic! Set the dial of the test receiver to the correct harmonic frequency, hook the probe over the output lead of the multiplier coil, and tune for maximum.

THE AC VOLTMETER FOR
GAIN CHECKS AND SIGNAL TRACING

The ac voltmeter can be a handy instrument. The standard rectifier-type voltmeter has a good frequency range, making it especially useful for gain checks and signal tracing in audio amplifiers.

Feed an audio signal—about 400 Hz—into the input of an audio amplifier. Put a blocking capacitor of any size from .01 to 0.1 μF in series with the meter. (Rectifier-type meters are affected by dc; the capacitor blocks any dc and lets you read only ac or signal voltages.) Now you can start at either end of the amplifier and trace the signal through each stage to find out whether there is gain or loss.

This test is handy in amplifiers that are weak but not dead. It's a rapid way to find the defective stage in a dead amplifier, too. Simply run through the circuit until you find the stage that has signal on the input but none on the output, and there you are.

This technique works with tubes or transistors and is especially useful in transistorized PC-board circuits. By using signal tracing first, you save the trouble of making voltage measurements and unsoldering parts before you have actually found the defective stage. The basic thing to check for here is small signal on the input (base or grid) and larger signal on the output (plate or collector), which would indicate that the stage does have gain. When you find a stage with no gain, or with a loss, that's the faulty one. ·

There is one exception to this rule in tube circuits and, to some extent, in transistors—the split-load phase inverter. As a rule, this stage is not designed to have any gain, but merely to divide and invert the signal. In the tube circuit, the load resistance is divided equally between the plate and cathode circuits. Thus, equal signal voltages appear on the two elements, but one signal is 180° out of phase with the other. There are several transistor circuits that do the same thing. To check for proper operation of such a circuit, measure the signal voltages on the input elements (grids or bases) of the following push-pull stage. The two signals should always be equal in amplitude. You can't check the phase without a scope, but chances are, if the signals are of equal amplitude, everything else is okay.

The ac voltmeter is also handy for finding troubles in stereo amplifiers. Feed the same audio signal into both inputs at the same time. Now measure the signal levels in corresponding stages of the two channels. At a given point, if you find one channel with a much lower signal than the other, that's the source of the trouble. Use the good channel as a guinea-pig to find out where the trouble is in the other channel.

The gain setting must be the same in both channels to prevent confusion. A good procedure is to turn both gain controls full on and then reduce the input signal level until the output level is about right. Don't overdrive; most stereo amplifiers have high gain and need only a small input signal. For the average phono input, 1 or 2 volts is plenty; for a microphone input, much less will do—.005 volt or less. Very high input signals can cause severe distortion or even damage transistors.

POWER OUTPUT TESTS
FOR PA AND HI-FI AMPLIFIERS

Check the power output of PA amplifiers *and* hi-fi's to see if that 30-watt amplifier is actually able to deliver 30 watts output. There is a simple test. After repairs have been completed and the amplifier is theoretically in first-class shape, hook up a properly matched load resistor across the output, feed in an audio signal, and read the power output by measuring the audio voltage across the load resistor. Ohm's law does the rest.

Figure 9-6 shows how the equipment is set up for this test procedure. You need a resistor that matches the output impedance of the amplifier and has a rating high enough to handle the power, with a safety factor. For a 30-watt amplifier, a 50-watt resistor is good. You can get such resistors from surplus stores at reasonable prices. Otherwise, make them up from stock values. For instance, PA and hi-fi tube amplifiers usually have output transformers tapped at 4, 8, 16, and 500 ohms. Five 75-ohm 10-watt resistors in parallel give 15 ohms at 50 watts (for equal resistors, power ratings are totaled), and this is close enough for the 16-ohm tap. Five 2,500-ohm resistors in parallel give 500 ohms, etc.

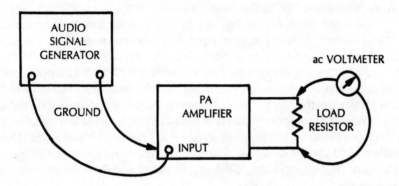

Fig. 9-6. This is the set-up for testing the power output of a hi-fi or PA amplifier.

With the load resistor hooked up, feed a low-level audio signal into the input; 1,000 Hz is a good frequency since most audio measurements are made at this frequency. Actually, the frequency doesn't make too much difference, as long as you're somewhere between 500 Hz and 5,000 or 6,000 Hz. (On most transistor amplifiers, especially the older ones, *don't* feed in a high-frequency signal—say, 15,000 Hz—at high power. The output transistors will overheat.) Remember this precaution: *Never* turn on an amplifier without the load resistor of the speaker hooked up. Even in tube amplifiers, you can burn out the output transformer in a very short time. And transistors can go in a fraction of a second if they are run without the proper load. Never short transistor outputs!

The service data gives the correct input level for many amplifiers. However, we're in the power output stage; can it deliver the rated power? To find out, hook an ac voltmeter across the load resistor and fire up the amplifier. If you're using a 15-ohm resistor and the amplifier is rated at 30 watts, for example, you should read at least 21 volts. Using $W = E^2/R$ and transposing to calculate E, you get $E^2 = 30 \times 15$, which yields about 21 volts for E.

This is also a good voltage-amplification or sensitivity check. For example, on a phono input you should get full power with the normal input level. With a high-output phono cartridge, this would be about a 2-volt input signal. On a low-output microphone, it would be about 5 mV, etc. If you can get full output only by overdriving the input to two or three times normal, then one of the voltage amplifier stages isn't giving enough gain.

10
Digital Circuits

Electronics circuits can be divided into two broad categories; analog and digital. These two types are fundamentally different. In the past, virtually all electronics circuits were analog. Almost all of the circuits discussed so far in this book are analog circuits.

But in the last decade or so, an electronics revolution has been taking place. Digital circuitry is becoming increasingly common, even in applications that were formerly solidly in the analog domain. Digital circuits are not used only in computers and calculators. They also show up in television sets, tape recorders, stereos, test equipment, alarm systems, and almost every conceivable type of electronics equipment.

Today's electronics technician has to be able to cope with digital circuitry. This is a frightening thought for many "traditional" technicians. Some technicians with long experience working with analog circuits feel intimidated by digital electronics.

There is really no need for such trepidation. Digital electronics *is* quite different from analog electronics in many ways, and servicing it often requires the technician to think in somewhat different ways. But there is considerable overlap between digital and analog circuitry. More importantly, digital electronics is inherently simpler in concept than analog electronics. If you can understand the workings of a switch, you can understand digital circuits. Digital circuits are simply combinations of various switching functions.

In a digital circuit, a signal may take on one of only two possible values. A LOW signal is usually just slightly above ground potential, and a HIGH signal is a little bit below the supply voltage. There are no other possibilities. These two states are given various names, but they all mean the same thing;

LOW	HIGH
0	1
NO	YES
FALSE	TRUE
OFF	ON

(In some specialized applications LOW is called "1" and HIGH is called "0". This is done for convenience, and nothing is changed except the arbitrary names used.)

There is no ambiguity in a digital circuit. The signal is either clearly LOW or clearly HIGH. If it is not definitely one state or the other, then something is unquestionably wrong. An analog signal, on the other hand, is inherently ambiguous. Say a signal at a certain point in the circuit is supposed to be five volts. Would 4.75 volts be acceptable? How about 5.5 volts? There is always a margin for error. In a digital circuit there is no margin for error. Only "yes's" and "no's" are permitted—"maybe's" can't exist.

Digital circuits are built around "semiconductors" called *gates* that are usually in IC form, so they can be treated as "black boxes". A digital gate produces a predictable output in response to the state of one or more inputs. The output/input pattern is frequently expressed in the form of a *truth table*. A truth table is simply a notation of what the output should be for every possible combination of inputs.

The simplest digital gate is a buffer. This is analogous to the buffer amplifier found in analog circuits. A buffer amplifier has a gain of one, so the output is the same as the input. Similarly, a buffer gate does not change the state of the signal passing through it. The truth table for a buffer looks like this;

INPUT	OUTPUT
0	0
1	1

Note that there is no other possible input to this device, so there can be no other output condition. The truth table covers *all* possibilities. The schematic symbol for a buffer is shown in Fig. 10-1.

INPUT	OUTPUT
0	0
1	1

Fig. 10-1. The simplest digital gate is the buffer.

An inverter works something like the inverting input of an op amp. The output always has the opposite state as the input. The schematic symbol and truth table for an inverter are given in Fig. 10-2. The letter with the overscore is read "A not."

INPUT	OUTPUT
A	$\overline{A}$
0	1
1	0

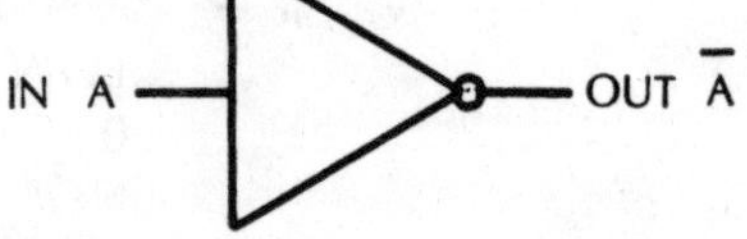

IN A ———▷o— OUT $\overline{A}$

Fig. 10-2. An inverter reverses the input state.

One-input gates are of limited use. More complex devices combine two or more inputs into one or more output(s). Multiple input gates might seem a little confusing at first, but they are easy enough to understand if you think about their names.

For example, a two-input AND gate is shown in Fig. 10-3. The output is HIGH *if and only if* input A *AND* input B are both HIGH. If either A or B (or both) is LOW, then the output must be LOW. This principle can be extended for any number of inputs. For example, Fig. 10-4 shows a four-input AND gate.

INPUTS	OUTPUT
A B	C
0 0	0
0 1	0
1 0	0
1 1	1

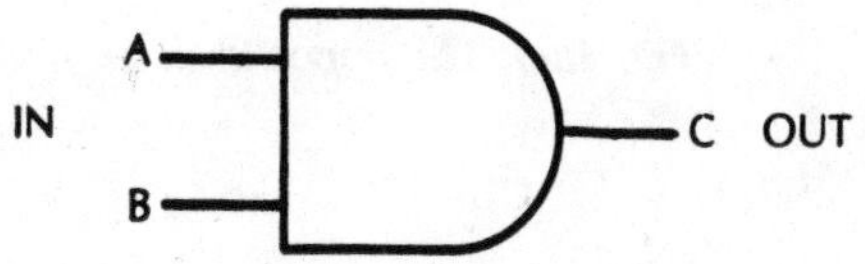

Fig. 10-3. This is the AND gate.

INPUTS	OUTPUT
A B C D	C
0 0 0 0	0
0 0 0 1	0
0 0 1 0	0
0 0 1 1	0
0 1 0 0	0
0 1 0 1	0
0 1 1 0	0
0 1 1 1	0
1 0 0 0	0
1 0 0 1	0
1 0 1 0	0
1 0 1 1	0
1 1 0 0	0
1 1 0 1	0
1 1 1 1	1

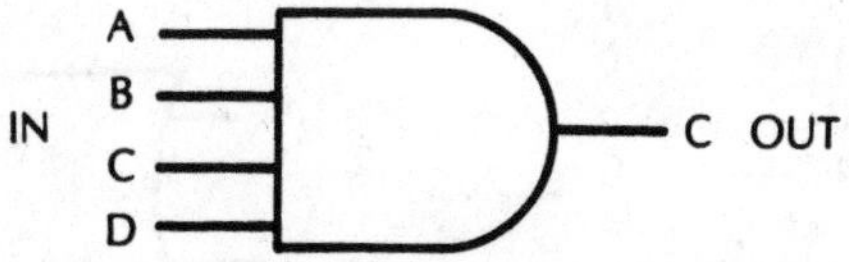

Fig. 10-4. Digital gates can have more than two inputs.

If you invert the output of an AND gate, as illustrated in Fig. 10-5, you get the opposite pattern. The output is HIGH *unless* both A and B are HIGH. This is called a NAND gate. The name is derived from "Not AND".

Another basic type of gate is the OR gate, shown in Fig. 10-6. The output is HIGH if either input A OR input B is HIGH. The output of an OR gate can be inverted, as shown in Fig. 10-7. This creates a NOR (Not OR) gate. The output is HIGH *if and only if* neither input A *NOR* input B is HIGH.

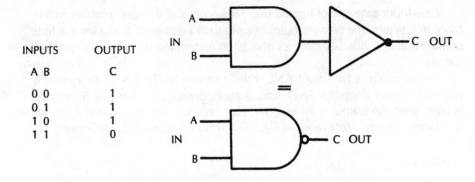

INPUTS	OUTPUT
A B	C
0 0	1
0 1	1
1 0	1
1 1	0

Fig. 10-5. The opposite of an AND gate is a NAND gate.

INPUTS	OUTPUT
A B	C
0 0	0
0 1	1
1 0	1
1 1	1

Fig. 10-6. Another common type of digital gate is the OR gate.

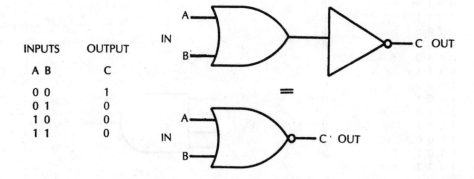

INPUTS	OUTPUT
A B	C
0 0	1
0 1	0
1 0	0
1 1	0

Fig. 10-7. Inverting an OR gate results in a NOR gate.

A variation of the basic OR gate is the X-OR, or eXclusive OR gate, illustrated in Fig. 10-8. The output is HIGH if *either* input is HIGH, but not *both*. The X-OR gate can be referred to as a difference detector, because the output goes HIGH only if the inputs are in different states. If the inputs are the same, the output will be LOW.

ICs to perform all of these basic gating functions are widely available and inexpensive. Any combination can be created by combining various gates. Digital

ICs that perform complex functions (such as counters, multiplexers, or even CPUs) contain multiple gate circuits internally. This discipline, *logic*, is the basis for computers.

INPUTS	OUTPUT
A B	C
0 0	0
0 1	1
1 0	1
1 1	0

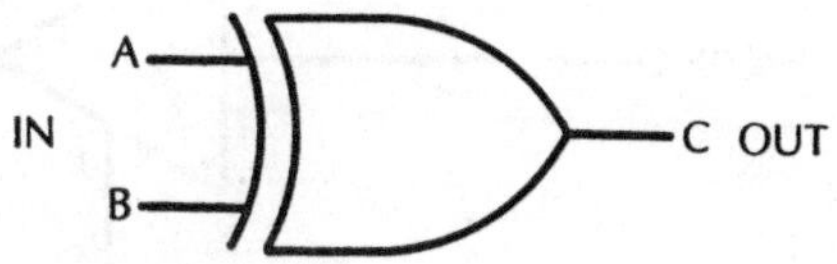

Fig. 10-8. The X-OR gate is a variation on the basic OR gate.

Space does not allow a discussion of all of the common digital ICs. Many books on the subject are available. One such volume is *Digital Interfacing with an Analog World* by Joe Carr (TAB Book #2850).

In servicing digital equipment, you need to know what signal level should be at a given point under certain circumstances. Essentially this is the same idea used in testing analog circuits. There are just fewer possible signal values in a digital circuit.

In most digital circuits, you see a small capacitor across the power supply leads of each IC. This capacitor's job is to filter out noise and brief transients in the supply lines. Digital ICs are very sensitive to power supply variations, even if they are very brief. If a piece of digital equipment starts behaving erratically, suspect problems in the supply voltage. Make sure the voltage is correct. Power supplies for digital circuitry should be very well regulated. Monitor the supply lines with an oscilloscope to see if there is excessive noise or transients. If just one IC seems to be misbehaving, the capacitor across its supply leads is a likely culprit. It could be open or leaky. Also make sure that there is a good connection to all of the IC's pins, especially if a socket is used.

LOGIC PROBES

The simplest type of digital test equipment is the logic probe. This is a device that indicates the current logic state (HIGH or LOW) at a specific point in the circuit. A super simple logic probe made from a single inverter section is shown in Fig. 10-9. The LED lights when the probe detects a LOW signal. If the LED does not light, the logic state is assumed to be HIGH. A ground connection must be made between the probe and the circuit being tested. The supply voltage for the probe can be tapped off from the power supply of the circuit under test.

While functional, this simple logic probe leaves a lot to be desired. There is no way to distinguish between a no-signal condition and a LOW logic state. Also, in many digital circuits, the logic state changes back and forth at a high rate, often at frequencies above 1 MHz. When the LED is lit, it could indicate a HIGH state or a stream of rapid pulses.

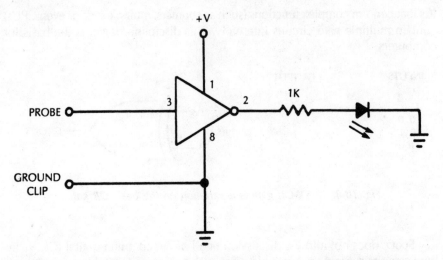

Fig. 10-9. A simple logic probe circuit. Pin numbers are for a CD4049 hex inverter IC.

An improved version of the circuit is illustrated in Fig. 10-10. This time there are two indicator LEDs. Together they can indicate four possible conditions;

LED 1	LED 2	INDICATED CONDITION
dark	dark	no signal
dark	lit	LOW
lit	dark	HIGH
lit	lit	pulses

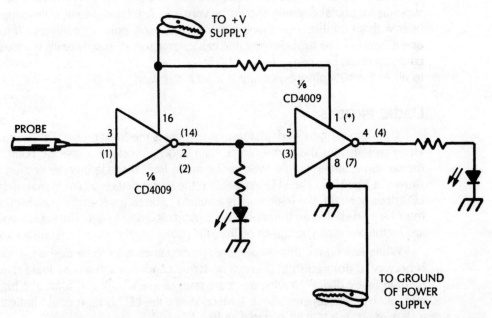

Fig. 10-10. This is an improved version of the logic probe circuit shown in Fig. 10-9.

MONITORING BRIEF DIGITAL SIGNALS

In many digital circuits, a signal might hold one state (HIGH or LOW) most of the time, but an occasional brief pulse to the opposite state will be critical to correct circuit operation. Such pulses can be extremely brief, perhaps only a tiny fraction of a second. The LED in a simple logic probe like those discussed in the preceding section would flash on and off too quickly to be visible to the human eye.

To detect such brief one-shot pulses, most commercial logic probes include a pulse stretcher circuit to extend the time the LED is lit. There are two basic approaches to pulse stretching. One is to use a timer in a monostable multivibrator mode. When the pulse is detected, it triggers the timer which lights the LED for a fixed period of time, regardless of how brief the triggering pulse is.

The other approach uses a flip-flop. A flip-flop (or bistable multivibrator) can hold either logic state indefinitely. Each time it is triggered, the flip-flop reverses its output state. When the flip-flop detects a pulse, the LED lights and stays lit until a second pulse is received or the circuit is manually reset (usually via a small push button).

SIGNAL TRACING IN A DIGITAL CIRCUIT

A logic probe is used in much the same manner as a signal tracer is used in an analog circuit. You can either start at the circuit's final output and work your way backwards through the circuit, or you can start at the input and work your way towards the output. In either case, when the signal is suddenly not what it should be, you have isolated the troublesome stage.

To perform signal tracing in a digital circuit, you need a very good understanding of what is supposed to be happening at each stage of the circuit. Work closely with the schematic and determine the function of each IC. If there is no signal at all, or if the output of a given IC remains constant regardless of the signals applied to any of its inputs, something is obviously wrong. No IC's function is to sit there and do nothing. Bad ICs will show one of these symptoms.

DIGITAL SIGNAL SHAPE

All digital circuits work with square waves (or rectangular waves). Most digital gates cannot reliably respond to other waveshapes. If you have a digital circuit that is behaving erratically, monitor the digital signals with an oscilloscope. You should get clean square wave without excessive ringing or noise. The tops and bottoms should be reasonably flat. The sides should be steep and straight. Distorted rise times or fall times can confuse many digital circuits.

POWER SUPPLY PROBLEMS IN DIGITAL CIRCUITS

The majority of problems in digital circuits seems to be traceable to troubles in the power supply. Digital ICs tend to be very power-supply sensitive. Some types are more sensitive than others. TTL-type digital ICs are cheap and durable, but the supply voltage for these devices must be between 4.75 and 5.25 volts.

CMOS devices (which are becoming the norm) will accept a wider range of supply voltages but are still sensitive to severe noise or ripples. Transients (brief bursts) in the power line can cause erratic operation or even permanent damage to some of the digital chips.

Be very critical when taking measurements in a digital power supply circuit. If anything seems questionable, treat it as if it is bad. Better safe than sorry.

Again, to protect digital ICs from transients and power supply noise, there is usually a small capacitor connected across the power supply pins of each individual IC. These capacitors are a fairly common source of trouble. If just one IC seems to be acting up, check the power bypass capacitor. It could be open or leaky.

In some inexpensive equipment, bypass capacitors cannot be used. If you want to guard against future problems, it might be worth your while to add the omitted capacitors yourself. They are certainly cheap enough and can usually be tack-soldered in place without too much difficulty. Be careful not to overheat the IC or you can destroy it. Always use a heat sink whenever you do any soldering near an IC. The capacitor should be mounted physically as close to the body of the IC as possible.

LOGIC ANALYZERS

A fairly new type of test instrument is the logic analyzer. This instrument can be thought of as the digital equivalent of an oscilloscope. It permits the technician to directly view the signals in a digital circuit.

In a nutshell, what the logic analyzer does is to sample digital signals and store them for later review. The stored signals are in the form of 1's and 0's.

A logic analyzer can be used in any digital circuit. It is most useful in microprocessor-based circuits. Generally the logic analyzer is used to monitor the signals appearing on the address or data buses. Several signals are monitored simultaneously.

Some logic analyzers display the stored data directly on an internal display. Others feed their output data to an ordinary oscilloscope for display.

The logic analyzer is really too complex to discuss in full detail here. It would take a book to completely cover its operation and use. If you do much work with digital (especially microprocessor-controlled) equipment, you should look into acquiring a logic analyzer.

PULSE GENERATORS

Pulse generators are becoming increasingly common on test benches. Basically, it is a special purpose signal generator. The circuitry is not dissimilar to the function generator.

A pulse generator puts out clean square waves and rectangular waves with steep rise times and fall times. The signals are suitable for use as test signals in test equipment, and a number of triggering modes are usually offered for various types of tests. A pulse generator can sometimes be used as a regular analog function generator, also.

Most of the better pulse generators offer a number of special features. One feature that can be useful for checking an erratic digital circuit is variable rise time/ fall time. Starting from the minimum settings (steepest sides on the waveform), the times are gradually increased until the device being tested starts showing the faulty symptoms. Now you know how much the signal can be distorted without trouble. Then use your scope to find the stage that distorts the pulse past the allowable amount.

A full discussion of the pulse generator is beyond the scope of this book, but if you do much work with digital circuits, a good pulse generator on your workbench is a handy tool.

Index

Edited by Lisa A. Doyle

Other Bestsellers of Related Interest

THE ILLUSTRATED DICTIONARY OF ELECTRONICS—5th Edition
—Rufus P. Turner and Stan Gibilisco

This completely revised and updated edition defines more than 27,000 practical electronics terms, acronyms, and abbreviations. Find up-to-date information on basic electronics, computers, mathematics, electricity, communications, and state-of-the-art applications—all discussed in a nontechnical style. The author also includes 360 new definitions and 125 illustrations and diagrams. 736 pages, 650 illustrations. Book No. 3345, $24.95 paperback only

GORDON McCOMB'S GADGETEER'S GOLDMINE!: 55 Space-Age Projects
—Gordon McComb

This exciting collection of electronic projects features experiments ranging from magnetic levitation and lasers to high-tech surveillance and digital communications. You'll find instructions for building such useful items as a fiberoptic communications link, a Geiger counter, a laser alarm system, and more. All designs have been thoroughly tested. Suggested alternative approaches, parts lists, sources, and components are also provided. 432 pages, 274 illustrations. Book No. 3360, $16.95 paperback only

THE ELECTRONIC PROJECT BUILDER'S REFERENCE: Designing and Modifying Circuits
—Josef Bernard

This unique guide is really three books in one: an electronic project book, a reference, and an idea book that shows you how to explore the world of electronics. Josef Bernard illustrates the principles of electronics design through projects, explaining why each component is necessary and how the required values are determined. He shows you how to build many different circuits, and then modify them for a multitude of uses. 208 pages, 108 illustrations. Book No. 3260, $16.95 paperback only

500 ELECTRONICS IC CIRCUITS WITH PRACTICAL APPLICATIONS—James A. Whitson

More than just an electronics book that provides circuit schematics or step-by-step projects, this complete sourcebook provides both practical electronics circuits AND the additional information you need about specific components. You will be able to use this guide to improve your IC circuit-building skills as well as become more familiar with some of the popular ICs. 336 pages, 600 illustrations. Book No. 2920, $29.95 hardcover, $24.95 paperback

GORDON McCOMB'S TIPS & TECHNIQUES FOR THE ELECTRONICS HOBBYIST
—Gordon McComb

This volume covers every facet of your electronics hobby from setting up a shop to making essential equipment. This single, concise handbook will answer almost all of your questions. You'll find general information on electronics practice, important formulas, tips on how to identify components, and more. You can use it as source for ideas, as a textbook on electronics techniques and procedures, and a databook on electronics formulas, functions, and components. 288 pages, 307 illustrations. Book No. 3485, $15.95 paperback only

HOW TO READ ELECTRONIC CIRCUIT DIAGRAMS—2nd Edition
—Robert M. Brown, Paul Lawrence, and James A. Whitson

In this updated edition of a classic handbook, the authors take an unhurried approach to the task. Basic electronic components and their schematic symbols are introduced early. More specialized components, such as transducers and indicating devices, follow—enabling you to learn how to use block diagrams and mechanical construction diagrams. Before you know it, you'll be able to identify schematics for amplifiers, oscillators, power supplies, radios, and televisions. 224 pages, 213 illustrations. Book No. 2880, $14.95 paperback only

**TROUBLESHOOTING AND REPAIRING
ELECTRONIC CIRCUITS—2nd Edition**
—Robert L. Goodman

Here are easy-to-follow, step-by-step instructions for troubleshooting and repairing all major brands of the latest electronic equipment, with hundreds of block diagrams, specs, and schematics to help you do the job right the first time. You will find expert advice and techniques for working with both old and new circuitry, including tube-type, transistor, IC, microprocessor, and analog and digital logic circuits. 320 pages, 236 illustrations. Book No. 3258, $21.95 hardcover only